T0185835

# Lecture Notes in Bioinformatics          **12099**

Subseries of Lecture Notes in Computer Science

More information about this series at http://www.springer.com/series/5381

Carlos Martín-Vide · Miguel A. Vega-Rodríguez ·
Travis Wheeler (Eds.)

# Algorithms for Computational Biology

7th International Conference, AlCoB 2020
Missoula, MT, USA, April 13–15, 2020
Proceedings

 Springer

*Editors*
Carlos Martín-Vide (iD)
Rovira i Virgili University
Tarragona, Spain

Miguel A. Vega-Rodríguez (iD)
University of Extremadura
Cáceres, Spain

Travis Wheeler (iD)
University of Montana
Missoula, MT, USA

ISSN 0302-9743          ISSN 1611-3349  (electronic)
Lecture Notes in Bioinformatics
ISBN 978-3-030-42265-3          ISBN 978-3-030-42266-0  (eBook)
https://doi.org/10.1007/978-3-030-42266-0

LNCS Sublibrary: SL8 – Bioinformatics

This Springer imprint is published by the registered company Springer Nature Switzerland AG
The registered company address is: Gewerbestrasse 11, 6330 Cham, Switzerland

# Preface

These proceedings contain the papers that should have been presented at the 7th International Conference on Algorithms for Computational Biology (AlCoB 2020) which was planned to be held in Missoula, Montana, USA, during April 13–15, 2020. The conference was postponed due to the coronavirus pandemic and will be merged with AlCoB 2021.

The scope of AlCoB includes topics of either theoretical or applied interest, namely:

- Sequence analysis
- Sequence alignment
- Sequence assembly
- Genome rearrangement
- Regulatory motif finding
- Phylogeny reconstruction
- Phylogeny comparison
- Structure prediction
- Compressive genomics
- Proteomics: molecular pathways, interaction networks, mass spectrometry analysis
- Transcriptomics: splicing variants, isoform inference and quantification, differential analysis
- Next-generation sequencing: population genomics, metagenomics, metatranscriptomics, epigenomics
- Genome CD architecture
- Microbiome analysis
- Cancer computational biology
- Systems biology

AlCoB 2020 received 24 submissions. Most papers were reviewed by three Program Committee members. There were also a few external reviewers consulted. After a thorough and vivid discussion phase, the committee decided to accept 15 papers (which represents an acceptance rate of about 62%). The conference program included three invited talks and some poster presentations of work in progress.

The excellent facilities provided by the EasyChair conference management system allowed us to deal with the submissions successfully and handle the preparation of these proceedings in time.

We would like to thank all invited speakers and authors for their contributions, the Program Committee and the external reviewers for their cooperation, and Springer for its very professional publishing work.

January 2020

Carlos Martín-Vide
Miguel A. Vega-Rodríguez
Travis Wheeler

# Organization

AlCoB 2020 was organized by the University of Montana, Missoula, USA, and the Institute for Research Development, Training and Advice (IRDTA), Brussels/London, Belgium/UK.

## Program Committee

| | |
|---|---|
| Mani Arumugam | University of Copenhagen, Denmark |
| Colin Dewey | University of Wisconsin–Madison, USA |
| Joe Felsenstein | University of Washington, USA |
| Olivier Gascuel | Pasteur Institute, France |
| Debashis Ghosh | University of Colorado, USA |
| Daniel Huson | University of Tübingen, Germany |
| Miriam Konkel | Clemson University, USA |
| Alla Lapidus | Saint Petersburg State University, Russia |
| Aron Marchler-Bauer | National Center for Biotechnology Information, USA |
| Maria-Jesus Martin | European Bioinformatics Institute, UK |
| Carlos Martín-Vide (Chair) | Rovira i Virgili University, Spain |
| David H. Mathews | University of Rochester, USA |
| Aaron McKenna | Dartmouth College, USA |
| Ryan E. Mills | University of Michigan, USA |
| Burkhard Morgenstern | University of Göttingen, Germany |
| Sayan Mukherjee | Duke University, USA |
| Houtan Noushmehr | Henry Ford Health System, USA |
| Knut Reinert | Free University of Berlin, Germany |
| Joel Rozowsky | Yale University, USA |
| Russell Schwartz | Carnegie Mellon University, USA |
| Temple F. Smith | Boston University, USA |
| James Taylor | Johns Hopkins University, USA |
| Zlatko Trajanoski | Medical University of Innsbruck, Austria |
| David A. Wheeler | Baylor College of Medicine, USA |
| Travis Wheeler | University of Montana, USA |
| Shibu Yooseph | University of Central Florida, USA |

## Additional Reviewers

Frederic Lemoine
Anna Zhukova

## Organizing Committee

| | |
|---|---|
| Sara Morales | IRDTA Brussels, Belgium |
| Manuel Parra-Royón | University of Granada, Spain |
| David Silva (Co-chair) | IRDTA London, UK |
| Miguel A. Vega-Rodríguez | University of Extremadura, Spain |
| Travis Wheeler (Co-chair) | University of Montana, USA |

# My Blue Whale: Seeking Order in a Chaotic World. An Autobiographical Reflection (Abstract of Invited Talk)

Tamar Schlick

Department of Chemistry and Courant Institute of Mathematical Sciences, New York University, 251 Mercer Street, New York, NY 1001

**Abstract.** This autobiographical reflection originated in an assignment for a Science Communication Workshop offered to ten selected science faculty at NYU last fall taught by science writer Stephen Hall. We were asked to write on the beginning of our journey in science. In my story, I recall my emerging interests in seeking order and patterns since I was a toddler growing up in Haifa, Israel. This journey eventually took me to the United States, where I earned degrees in Applied Mathematics and pursued research work in mathematical biology and computational biophysics. This research, described in my Keynote lecture on folding genes at the 7th International Conference on Algorithms for Computational Biology (AlCoB 2020) in Missoula, Montana in April, blends computational mathematics, biology, chemistry, physics, computer science, and engineering in a creative way.

## Leviatan

I'm not sure exactly when I decided to become a scientist but, looking back, a desire to seek pattern and order was there since I was a toddler, growing up in the seaside town of Haifa, Israel.

When I was three years old, my parents gave me a big blue plastic fish on yellow wheels. It had a string at the mouth and a cover that opened, so I could fill the fish's belly. I coined this toy my "whale" ("Leviatan" in Hebrew), as its size was overwhelming.

I promptly started a habit that intrigued my puzzled parents. I would load Leviatan with all my treasures: wooden blocks, lego pieces, small dolls, and coloring set, and noisily wheel my whale along the main corridor of our apartment—from bathroom to front door. Arriving at the front door, I would unload all my loot, line up the pieces like wooden soldiers and, only when satisfied, dismantle my workmanship, load my Leviatan again, and proceed back to the bathroom. I repeated these journeys without tiring for hours.

As soon as I started to attend elementary school, I wanted to emulate the teaching experience at home, but with the roles reversed. In my little bedroom, I would line up

---

**Dedication**: I dedicate this essay to my mother, Dr. Shulamith Schlick, who passed away on Jan. 15, 2020. Her dedication to science, pursuit for perfection, and intellectual curiosity have been an inspiration.

**Fig. 1** A photograph of my blue whale.

all my dolls and beloved teddy bear (still with me today) in three neat lines, assign numbers and names to my students, and proceed to teach them arithmetic and reading/writing in front of my closet, using a long ruler as my pointer. My father, relieved that the whale migrations were on hold, decided to indulge my teaching fantasy by painting my closet with blackboard paint. He also gave me a set of colored chalks to complete the mission. As soon as I came back from school, I would rush to line up my diligent students on the floor in their assigned seats and proceed to lecture them on whatever I learned in school that day. When my father would open the door and enter my room to call me for dinner, I would turn to him with great embarrassment and say quietly, so that my students wouldn't hear: "Dad, can't you see that we are in the middle of class"?

## Anne Frank's Story

A year later, my parents took me to the play "The Diary of Anne Frank" at the Technion Institute of Technology theatre, which presented many children's educational programs. Although Anne's plight was remote from my comfortable life in the sleepy city of Haifa, far in those days from bustling Tel Aviv, growing up in Israel instilled a feeling of fear and uncertainty about the world, where enemies are real and war preparedness is part of life. I was so shaken and inspired by the story of Anne and her family in Holland during World War II that I decided on the spot to start writing a diary. I now have three meters of diaries since that day, as I haven't stopped writing since. As I read and re-read my entries and look at the large collection of drawings, charts, photos, and article scraps I have amassed in these volumes, many events and memories come alive that have long escaped me.

**Fig. 2** Teacher in training.

## Data Collection and Analysis

From my first simple notebook which I covered with the red diamond shelf-liner paper I found in our kitchen, I loved to illustrate my writings with maps, calculations, photos, newspaper cutouts, and precise tidbits of information. In one of the first entries, I discuss the high price of entry to the private swimming pool near our home where I learned to swim. This was the beautiful outdoor pool of the Dan Carmel hotel, a sister to the well known five-star King David hotel in Jerusalem. I reported in my diary the prices of entry (in the lira currency of those days) for adults and children, on weekdays and on holidays; I then compared the total cost for our family to that required for swimming at a nearby public pool. These habits of finding value and efficiency still resonate for me today.

Whenever I received report cards, I prepared tables in my diary, recorded the grades and comments I received from teachers, and then compared these to the older sets. As those lists got longer and more complicated, I found ways of recording that information in new ways: by changing the page orientation, taping added pages to widen the margins, and finding more compact ways to label and tabulate the data. This excitement about data collection and comparative analysis rings true today as I pursue my research in biophysical modeling.

My parents, a scientist and an engineer, were very busy during my school years, proving themselves at their jobs and working hard to support our growing family. Yet they sent me to an excellent private school and made sure to enroll me in many extracurricular classes, from swimming to piano lessons. I inherited from them many math books and tools. I loved to play with these gadgets, like protractor, triangular rulers, and ellipse drawing instruments of all shapes and sizes. These studies of patterns and connections are not unlike the search for chromosome rearrangements and folding that occupy my research group today.

## Puzzles and Patterns

I liked solving puzzles and building objects. I had subscriptions to magazines with puzzles and challenges and solved every one in order. I liked story problems that most of my classmates despised, as they preferred numerical calculations. A special event for me was a bimonthly radio show for adults called the "Island of Treasure." In this contest, a variety of clues were provided, and contestants had to decipher and manipulate them to arrive at a specific location in Israel where the prize was hidden. The radio show often extended late into the night, and I would lie in bed curled up with my small transistor radio and follow the treasure hunt, trying to piece together the difficult historical and geographical clues.

In middle school, I especially loved studying maps in geography class, drawing routes and paths from our small country to far away places, and learning about the weather, landscape, and culture across the world. No wonder I became fascinated by graph theory to study biological systems decades later! We use graph theory extensively now in my group to represent, analyze, and design novel RNA molecules.

When we were assigned a research project on a country of our choice, I chose Canada and wrote to the Canadian embassy in Israel as well as to my uncle in Montreal. I poured over the beautiful colorful maps and pictures I soon received, excited to see pretty snow and peaceful landscapes, so foreign and distant from my chaotic homeland, where war and death were well known to children. Despite the sheltering by my parents to many realities, the volatility of the political climate penetrated my life, where fathers, uncles, and brothers went to war and where army draft awaited us all immediately after high school. There was always a sense that we have to live life to the fullest, seizing every moment, before it may be too late. I still live with that missive, trying to make every minute productive.

## Proving Hypotheses

In high school, I chose a math/physics track because a teacher recommended it. I loved all subjects, without much discrimination. Whether determining the answer of a mathematical puzzle or finding the origin of a lost tribe, I found the process of establishing knowledge fascinating.

Although I had good rapport with many of my high-school teachers, one of them infuriated me because he seemed to grade our homework without rhyme or reason. I developed a hypothesis, which I shared with my classmates, that this teacher gave all the pretty girls who smiled at him an A grade, while the rest of us received a B or lower depending on our gender and level of our enthusiasm for his teaching style and content, both of which I despised. Determined to prove my conjecture, I organized the class to submit homework assignments where half a dozen of us had filled the pages with nonsensical content, namely repetition of the sentence "The cow chewed green grass in the pastures." Sadly, my hypothesis was quickly proven to be true, as the teacher had not even noticed the garbage content and awarded us the usual grades. My complaint, which I promptly took to the principal with the evidence, eventually resulted in this teacher being expelled due to this and other incriminating incidents. This hypothesis-driven proof energized my righteous spirit and emboldened my interest in scientific research.

Fortunately, when my high-school physics teacher became interested in my progress, I started to think more seriously about what I wanted to study. He suggested that I learn as much science and math as possible and continue an academic rather than professional track. Numbers and patterns had a special resonance to me, and I decided to pursue my interest in data collection/analysis and problem solving more systematically.

## Applied Math and Theoretical Chemistry Research

Soon, after a combination of world events and personal family circumstances, my family immigrated to the United States and settled in a Michigan suburb north of Detroit. With a four-year Merit scholarship to attend Wayne State University, I connected with a small group of honor students who were almost exclusively set on a pre-med academic track. For a while, I entertained going into medicine as well. However, at the wise advice of a counselor, I promptly found a summer job at a local hospital and was soon turned off by the vials of blood and loud, cluttered environment. The organized and quiet world of numbers and theorems, with concepts and logical deductions that I could explore on my own seemed much more appealing and suitable to my personality. Chemistry Professor Bill Hase gave me the opportunity to work as a researcher in his lab, allowing me to gain invaluable experience in computer modeling and theoretical chemistry research. I enjoyed this research experience throughout my undergraduate years.

In tandem with the lab research experience, I took as many math classes as possible, solving hard homework problems and take-home exams with my fellow students, mostly men. I saw myself as one of them and so did they. As I enjoyed more advanced math courses, notably group theory and real analysis, I realized how little the public knows about math. Many people associate mathematics with arithmetic, with tasks like calculating compound interest rates or adjusting recipe quantities. Higher-order math, however, is about abstract thought and ideas, concepts to organize numbers and algorithms, and discovery of new paths and patterns by deduction and resourcefulness. Such logic and creativity excited me. From real to complex variables, graph theory to number theory, numerical analysis to computational mathematics, I found applied math infinitely wondrous. Rather than seek exact solutions to made-up problems, applied mathematicians solve real-world problems approximately. These imperfectly phrased but relevant problems are solved by creative theoretical analysis and computer modeling.

## Mathematical Biology and Chemical Physics

My love for applied math soon took me through graduate school at NYU's Courant Institute of Mathematical Sciences, working with Professors Charles Peskin from Mathematics, Michael Overton from Computer Science, and Suse Broyde from

Biology. I chose for my thesis topic modeling the DNA molecule and recruited these advisors to guide me in this difficult task. Solving biological problems using mathematical tools and computer modeling seemed like a perfect marriage. This research combined my interest in both life and mathematical sciences with creative problem solving and scientific discovery.

Upon graduation with a doctoral degree in applied mathematics, my academic path swiveled back to Israel on a postdoctoral fellowship at the Weizmann Institute of Science. There I learned from a grand pioneer in chemical physics, the late Shneior Lifson, more about modeling and molecular mechanics. These combined interdisciplinary experiences eventually led me back to New York City, where I work on modeling and simulating biological systems like DNA and educating a future generation of computational biologists.

## Computational Biophysics and Genome Organization

In this exciting multidisciplinary field, we decipher biological phenomena by computer modeling and simulation. In the subfield of genome organization on which my group focuses, we seek to understand how the molecular agent of heredity, DNA (deoxyribose nucleic acid), folds and unfolds in our cells to regulate basic processes of life. These processes include DNA replication, expression of genetic traits, and disease progression. Human DNA is organized in our cell nuclei as chromosomes in the shape of long spaghetti-like string called the chromatin fiber. This spaghetti adopts different shapes and separates into various compartments in our cells depending on the state of the cell. Understanding these shapes and patterns is key to understanding regulation of the basic processes of life. While the first order of biological information is encoded in our DNA through the DNA (primary base) composition (those A,C,G,T components), another level of regulation distinct from the sequence is expressed through the chromatin fiber structure. This *"epigenetic regulation"*—through changes in the structure of this spaghetti-like string—is now believed to hold part of the key for understanding human disease. Our models for the chromatin fiber are revealing folding patterns encoded in the chromatin fiber that direct the folding of genes, thereby determining when genes are silenced (not expressed) or when they are switched on (expressed). Knowing how to control this switch can help halt the progression of events that lead to abnormal cell growth in cancer. Such an understanding has immediate applications to disease diagnostics and therapy. On a technical level, the work involves developing innovative models and algorithms for simulating the structures and motions of these complex biological systems, merging mathematics, biology, chemistry, physics, computer science, and engineering ideas.

Whether I'm writing papers on chromosome folding or teaching at a summer school on cancer genomics, I still feel like that toddler, shuffling objects and ideas from place to place, trying to establish pattern and order and make discoveries to comprehend this complex and chaotic world. My Leviatan is alive and well, occupying a prime spot in our Greenwich Village apartment, having weathered the years a bit better than me. His reassuring presence reminds me of the far away sea and vast knowledge still awaiting discovery.

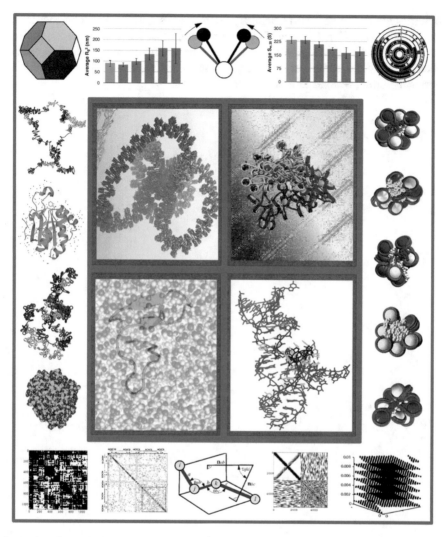

**Fig. 3** A collage of various plots, analyses, and images required in computational biophysics to study biological molecules and simulate their structures and motions to infer biological activity.

## Acknowledgments

I thank my family for making my journey in life and science a happy one. I thank my mentors and mentees for making this wonderful life in science possible. Current research is made possible through an NIH MIRA Award from the National Institute of General Medical Sciences R35GM122562. I thank Stephanie Portillo-Ledesma for assistance with two figures.

**Tamar Schlick** is Professor of Chemistry, Mathematics, and Computer Science at New York University. Her research focuses on computational biophysics, specifically chromosome folding and gene regulation. She has authored more than 200 research articles and is the author of an interdisciplinary textbook on molecular modeling (Springer-Verlag, Second Edition, 2010).

# Contents

## RNA-Seq and Other Biological Processes

# Genomics

# Parallel Generalized Suffix Tree
# Construction for Genomic Data

Md Momin Al Aziz$^{(\boxtimes)}$ ⓘ, Parimala Thulasiraman, and Noman Mohammed ⓘ

Computer Science, University of Manitoba, Winnipeg, Canada
{azizmma,thulasir,noman}@cs.umanitoba.ca

**Abstract.** After a decade of digitization and technological advancements, we have an abundance of usable genomic data, which provide unique insights into our well-being. However, such datasets are large in volume, and retrieving meaningful information from them is often challenging. Hence, different indexing techniques and data structures have been proposed to handle such a massive scale of data. We utilize one such technique: Generalized Suffix Tree (GST). In this paper, we introduce an efficient parallel generalized suffix tree construction algorithm that is scalable for arbitrary genomic datasets. Our construction mechanism employs shared and distributed memory architecture collectively while not posing any fixed, prior memory requirement as it uses external memory (disks). Our experimental results show that our proposed architecture offers around 4-times speedup with respect to the sequential algorithm with only 16 parallel processors. The experiments on different datasets and parameters also exhibit the scalability of the execution time. In addition, we utilize different string queries and demonstrate their execution time on such tree structure, illustrating the efficacy and usability of GST for genomic data.

**Keywords:** Parallel generalized suffix tree · Genomic data indexing · Parallel computation on genomic data · GST construction on disks

## 1 Introduction

The achievements in human genomics have been remarkable during the last decade. Concepts like genomic or personalized medicine and genetic engineering are slowly becoming reality which seemed impossible a few years ago. We are now capable of storing thousands of genome sequences from patients along with their medical records. Today, medical professionals utilize this large-scale data to study associations or susceptibility to certain diseases.

The recruitment for different genomic research is also increasing as the genome sequencing cost is ever-reducing through technological breakthroughs in the last few years. This growth in genomic data has resulted in consumer products where companies offer healthcare solutions and ancestry search based on human genomic data (*e.g.*, Ancestry.com, 23AndMe). Interestingly, all these

© Springer Nature Switzerland AG 2020
C. Martín-Vide et al. (Eds.): AlCoB 2020, LNBI 12099, pp. 3–15, 2020.
https://doi.org/10.1007/978-3-030-42266-0_1

applications share one major operation: *String Search*. Informally, the string search denotes the presence (and locations) of an arbitrary query nucleotide sequence in a larger dataset. The search results comprise the individuals who carry the same nucleotides in the corresponding positions. Thus, we can perceive the relation between an unknown sequence to pre-existing sequences with such search queries.

On the other hand, suffix tree is proven useful for searching different patterns or arbitrary queries on genomic data [2]. However, their construction suffers from the locality of reference as reported in the initial works [7]. This problem refers to the memory accesses in the same locations within a short period while building the suffix tree. Moreover, it gets severe as suffix trees perform best when the tree (vertices and edges) completely fits in the main memory. Unfortunately, this is quite impossible with off-the-shelf implementations and large scale genomic data.

In this work, we construct generalized suffix trees (GSTs) in parallel. There has been several attempts in efficient, parallel *suffix tree* building which considers only one sequence whereas GSTs represent multiple sequences [5]. We employed two different memory architectures for our parallel GST construction: (a) distributed and (b) shared memory. In a distributed architecture, we utilized multiple machines with completely *separate main memory* system interconnected within a network. On the contrary, these processors have several cores, which share the *same main memory*. These cores are employed in our shared memory model. Furthermore, we employ a data specific parallelism based on the fixed nucleotide set in this construction for the shared memory architecture. Finally, our GSTs are built on file system to remove the dependency for a sizeable memory requirement. We can summarize our contributions below:

- The primary contribution of this paper is a parallel framework using the distributed and shared memory models to construct GST for a genomic dataset.
- We also utilize the external memory (or disks) since GSTs for large-scale genomic data require notable memory size, which is usually not available in a single machine.
- We test the efficiency of our GST with multiple string search queries. Furthermore, we analyze the parallel speedup in terms of dataset size, number of processors, and components of the hybrid memory architecture.
- Experimental results show that we can achieve around *4.7 times* speedup compared to the sequential algorithm with 16 processors to construct the GST for a dataset with $n = 1000$ sequences and 1000 nucleotides each.

Notably, Ukkonen's algorithm [14] went out of memory for $n, m = 10,000$ dataset whereas our proposed approach takes 77.3 s with 16 processors.

## 2  Preliminaries

### 2.1  Haplotype Data

In this paper, we utilize the bi-allelic genomic data where the ratio of different nucleotides in a specific position of a chromosome is known beforehand. It is also called haplotype data, where each allele (or position) on the chromosome is inherited from a single parent. In other words, in one specific location, we can only perceive two variations for such a dataset; therefore, we utilize a binary representation. However, our proposed method is not limited to such binary representation and generalizable over any dataset with a fixed character set.

## 2.2   Generalized Suffix Tree

**Suffix Trie and Tree:** Trie (from retrieval) is a data structure where each element of the data are placed in the vertex of a tree. Here, the edges represent the relation of one data to the other. In our problem scenario, each nucleotide of the sequences can be seen as the other data points or vertices of a Trie.

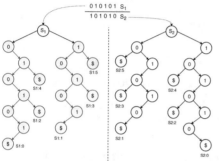 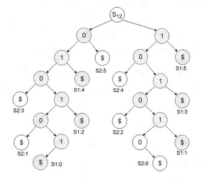

**Fig. 1.** Uncompressed Suffix Tree (Trie) construction

**Fig. 2.** GST from Fig. 1 where gray and white vertices are from S1 and S2, respectively

Suffix Trees are a compressed version of their Trie counterpart. For example, if a single vertex has only one child on a suffix trie, they are joined and denoted as a single vertex on the Suffix Tree. In Fig. 1, we show two suffix tries S1 and S2 from sequences 010101 and 101010 respectively. For any sequence S1 = 010101, we consider all possible suffixes such as $\{1, 01, 101, 01010, 10101, 010101\}$ and construct a trie. The simple approach to construct such a tree will require an iteration over all suffixes and add/merge new vertices if required.

The suffix tree will also represent the end of the sequence with a special end character ($). For example, the suffix 01 (left Fig. 1-S1) has an end character with label S1:4 which denotes the sequence number and the start position of the suffix. Formally, we use vertex label Sx:y $s.t.$ $x \in \{1, n\}$ and $y \in \{0, m - 1\}$ for $n$ sequences of $m$ length.

*Generalized Suffix Tree (GST):* Generalized Suffix Tree is a collection of suffix trees for multiple sequences. Here, we merge two suffix trees S1 and S2 from Fig. 1 and construct S12 in Fig. 2. Fundamentally, there are no difference in constructing GST as we need to build individual suffix tree per sequence and merge them afterwards. Thus, the runtime for one GST construction depends on these suffix tree construction and size. For example, the traditional algorithm (Ukkonen) to build the suffix tree has a linear runtime $O(m)$ for $m$ length sequences [14]. Therefore, $n$ sequences with $m$ characters will be at least $O(nm)$ along with the additional linear tree merging cost of $O(n)$.

### 2.3   Utility Measure

We considered different queries to test the utility of a GST. However, for brevity, we only report three problem instances which are related and incrementally challenging. The input will be a genomic dataset $D$ consisting $n$ individuals with $m$ nucleotides. Since we are considering $n \times m$ haplotypes, $D$ will have $\{s_1, \ldots s_n\}$ records with $s_i \in [0,1]^m$. The query can be of arbitrary length $(1 \leq |q| \leq m)$.

**Query 1 (Exact Match-EM).** For a genomic dataset $D$ and an arbitrary query $q$, an exact match will only return the records $x_i$ which observe $q[0, |q| - 1] = x_i[j_1, j_2]$ where $q[0, |q| - 1]$ denotes the full query and $x_i[j_1, j_2]$ is a substring of the record $x_i$ given $j_2 \geq j_1$ and $j_2 - j_1 = |q| - 1$.

**Query 2 (Set Maximal Match-SMM).** Similarly, for the same inputs, a set maximal match will return the record $x_i$ which have the following conditions:

1. $q[j_1, j_2] = x_i[j_1, j_2]$ where $j_2 > j_1$ (same length and positions),
2. $q[j_1 - 1, j_2] \neq x_i[j_1 - 1, j_2]$ or $q[j_1, j_2 + 1] \neq x_i[j_1, j_2 + 1]$, and
3. No other records $x_i'$ with substring $[j_1', j_2']$ where $j_2' - j_1' > j_2 - j_1$.

**Query 3 (Position-variant Set Maximal Match-PVSMM).** Finally, for a threshold $t$, the PVSMM will report all records where which follow:

1. $q[j_1, j_2] = x_i[j_1', j_2']$ where $j_2 - j_1 = j_2' - j_1' \geq t$,
2. $q[j_1 - 1, j_2] \neq x_i[j_1' - 1, j_2']$ or $q[j_1, j_2 + 1] \neq x_i[j_1', j_2' + 1]$, and
3. No other records $x_i''$ with substring $[j_1'', j_2'']$ where $j_2'' - j_1'' > j_2' - j_1'$.

## 3   Methodology

### 3.1   Architecture and Design Goals

The outline of the parallel generalized suffix tree construction and corresponding computation is depicted in Fig. 3. Here, the genomic dataset $D_{|n \times m|}$ is operated by the data owner and the researchers have $q$ queries on $D$. The researcher does not have any substantial processing power compared to the data owner since s/he is only interested in a minuscule portion of $D$.

The high-level design of our architecture is illustrated in Fig. 3, where the data is evenly partitioned between different computing nodes (in one/multiple clusters). Here, we consider and utilize two type of memory environment: distributed and shared. In the distributed memory, the machines are connected via network as they have *mutli-core* processors and their own physical memory (RAM). The mutli-core processors on these machines collectively use the physical memory which is called as shared memory. Hence, we have $|p|$ computing nodes which construct our desired GST jointly.

Our memory dispersion tackles one of the major disadvantages of the GST construction: the sizeable memory requirement for longer sequences. For example, a thousand length sequence can create a maximum of a thousand vertices, and $n$ sequences can lead to an order of $nm$. Thus, for an arbitrary genomic dataset, it often outruns the memory. Hence, this motivates us to construct our targeted GST in a *distributed memory setting*.

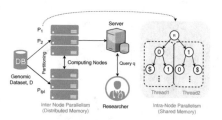

**Fig. 3.** Computational architecture where data owner has the dataset $D$ while a researcher submits query $q$

This leads to our proposed design where we distribute the data (partition) and build the suffix tree separately in different computing nodes. These nodes can construct each subtree which is later shared to the other nodes. These shared subtrees are then merged, and the final tree includes all suffix subtrees combining the outputs from all computing nodes (Sect. 3.2.4). The multiple processors in each node will also use shared memory model while constructing and merging their individual GST in parallel (Sects. 3.2.2 and 3.2.3). Therefore our three design goals can be summarized as follows:

1. Partition the dataset for different nodes in a *distributed memory architecture* where individual computing nodes receive a part of the data and only constructs a subtree of the final GST (Inter-node Parallelism)
2. As these nodes are equipped with multiple cores, they will build the individual GSTs in parallel using *shared memory architecture* (Intra-node Parallelism)
3. Use external memory to store and share the resulting GSTs to reduce sizeable main memory requirement.

## 3.2 Parallel GST Construction

### 3.2.1 Data Partitioning

We utilize different data partitioning scheme based on the memory locality, availability and the number of computing nodes:

**Horizontal** partitioning groups a number of sequences for the existing computing nodes. Each node will receive one such group and construct the corresponding GST afterwards. For example, if we have $n = 100$ sequences and $p = 4$ nodes, then we will split the data into 4 groups where each group will contain $|n_i| = 25$ records or genomic sequences. Each node, $p_i$ will build their GST on $|n_i|$ sequences of $m$ length in parallel without any communication. Figure 1 depicts a simple case of this partition scheme for $n = p = 2$.

**Vertical** partitioning divides the data across the columns and distributes it following the aforementioned mechanism. However, this scheme will have some additional implications while merging the resulting subtrees (Sect. 3.2.2). For example, if we have genomic data of length $m = 100$ and $p = 4$, we will have $n \times m_i$ partitions where each dataset will have $|m_i| = 25$ columns.

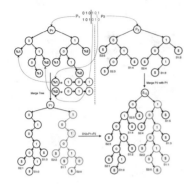
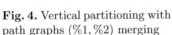
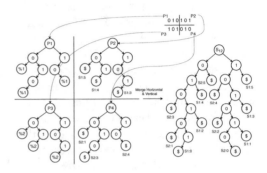

**Fig. 4.** Vertical partitioning with path graphs (%1, %2) merging

**Fig. 5.** Example of Bi-Directional partitioning scheme

**Bi-directional** data partitioning combines both the horizontal and vertical approach as it divides the data into both directions. Notably, it can only operate for $p \geq 4$ cases. Given $n = 100$, $m = 100$ and $p = 4$, each node will receive a $n_i \times m_i = 50 \times 50$ sized data for their computations.

### 3.2.2 Distributed Memory

We use several machines (or nodes) to build the final GST in parallel (*Inter*-node Parallelism), each with their individual global memory and connected via network. After receiving the partitioned data, these computing nodes are required to build their own GSTs. For example, if there are $p_0, \ldots, p_{|p|}$ nodes then we will have $GST_0, \ldots, GST_{|p|}$ trees. Regardless of the partitioning mechanism, we use the same linear time method to construct the suffix trees using Ukkonen's algorithm [14]. After these individual nodes build their GSTs, they need to share them for the merging operation described next.

In Fig. 1, we see a horizontally partitioned GST construction. Here, two suffix trees are merged where the grey and white colored nodes belonged to different trees. Notably, the merge operation did not duplicate any node at a particular depth. For example, if there was already a node with the value 0 is present, then it will not create another node and simply merge onto its branches. This condition is applied to all merge operations to avoid duplicate branches.

However, for vertical and bi-directional partitioning, the merging requires an additional step for datasets where $m_i < m$. We illustrate this in Fig. 4 where $n = 2, p = 2, m = 6$ and we are creating GST for the sequences S1,S2 = {010101, 101010}. Here, $p1$ operates on {010, 101} partitions whereas $p2$ generates the tree for {101, 010}. Here, the GST from $p1$ needs to have different end characters compared to $p2$ as each end points in a suffix tree needs to represent that the suffix has ended there. However, since we are splitting the data on columns, it needs to address the missing suffices.

Therefore, we perform a simple merge for all cases with $m_i < m$ as we add the *Path Graphs* with $m - m_i$ characters on the resulting GST. For example, in

Fig. 4, we add the path graph of 101 (represented as %1) in all end characters of S1 in $p1$ (after 010). Similarly, we need to add 010 for S2 represented as %2. During this merge, we also do not create any duplicate nodes. The addition of these paths will require the merge operation described next.

### 3.2.3   Shared Memory

In the distributed memory environment, the individual machines get a partition of the genomic data to build the corresponding GSTs. However, these machines or nodes also have multiple cores available in their respective processor which share the fixed global memory. Therefore, we also employ these cores to build and merge the GST in parallel.

We utilize the fixed alphabet size property of genomic data in our shared memory model (*Intra*-node Parallelism). Since there can be only two possible children of the root $(0/1)$, we can initiate two processes where one process will handle the 0 leading suffixes whereas the other process will operate on 1's. For example, in Fig. 1 two processes $p1$ and $p2$ will generate the suffix tree of $\{01, 0101, 010101\}$ and $\{1, 101, 10101\}$ respectively. The output will be two suffix trees, one from each process which can be joined with the root for the final tree.

It is noteworthy that the GSTs on the partitions can also be build with this shared environment. Here, we will partition the data into the cores and they will build, merge the GST in parallel. However, the number of cores and memory is limited in a non-distributed setting which will restrict larger GSTs.

### 3.2.4   Merging GSTs

As mentioned earlier, the merge operation takes two different GSTs and adds all their vertices. Hence, all $|p|$ GSTs are merged into the final $GST$ where $GST = GST_0 + \ldots + GST_{|p|}$. Here, we employed the shared memory parallelism as the children of the root $(0/1)$ are totally separate and do not have any common edges. In other words, we can treat the root's 0 branch separately from child 1. This allows us to perform the merge operation in parallel and utilize the Intra-node parallelism in each computing nodes.

Notably, merging one branch of a tree is a serial operation as multiple threads cannot add/update branches simultaneously. This creates a bottleneck as we need to perform merge operation in all $GST_i$'s and add the path graphs mentioned in Sect. 3.2.2. However, we can use multiple cores for different branches as mentioned in Sect. 3.2.3. For example, we can create two processes for handling the 0 and 1 branch from the root. This can be extended for the suffixes starting with $00, 01, 10, 11$ as well. Notably, this parallel operation can be followed for any dataset with fixed character set.

The full merge operation is depicted in Fig. 5 where we perform the bidirectional partition and merge accordingly. Inherently, the bidirectional strategy employs both vertical and horizontal merging strategies as the end columns do not include the $m - m_i$ characters.

**Table 1.** Horizontal and Vertical partition scheme execution time (in minutes) to build GSTs with number of processors $p = \{1, 2, 4, 8, 16\}$

| Data | Serial | Distributed | | | | Shared | | | | Hybrid | | | |
|---|---|---|---|---|---|---|---|---|---|---|---|---|---|
| | 1 | 2 | 4 | 8 | 16 | 2 | 4 | 8 | 16 | 2 | 4 | 8 | 16 |
| Horizontal partitioning | | | | | | | | | | | | | |
| 200 | 0.08 | 0.23 | 0.09 | 0.09 | 0.10 | 0.14 | 0.05 | 0.04 | 0.03 | 0.14 | 0.07 | 0.05 | 0.05 |
| 300 | 0.27 | 1.04 | 0.23 | 0.2 | 0.23 | 0.38 | 0.15 | 0.11 | 0.08 | 0.37 | 0.16 | 0.12 | 0.12 |
| 400 | 0.59 | 2.03 | 0.55 | 0.38 | 0.38 | 1.18 | 0.35 | 0.21 | 0.2 | 1.12 | 0.31 | 0.23 | 0.25 |
| 500 | 1.53 | 3.14 | 1.32 | 1.06 | 1.01 | 2.27 | 0.57 | 0.36 | 0.28 | 2.09 | 0.52 | 0.38 | 0.41 |
| 1000 | **14.55** | 16.23 | 8.34 | 6.31 | **6.09** | 17.38 | 5.56 | 3.27 | **2.28** | 17.14 | 4.18 | 3.12 | **3.08** |
| Vertical partitioning | | | | | | | | | | | | | |
| 200 | 0.08 | 0.19 | 0.08 | 0.05 | 0.03 | 0.16 | 0.07 | 0.04 | 0.02 | 0.14 | 0.05 | 0.03 | 0.02 |
| 300 | 0.27 | 0.56 | 0.28 | 0.17 | 0.09 | 0.48 | 0.22 | 0.16 | 0.08 | 0.39 | 0.13 | 0.10 | 0.06 |
| 400 | 0.59 | 1.41 | 1.05 | 0.36 | 0.16 | 1.44 | 1.01 | 0.34 | 0.19 | 1.21 | 0.32 | 0.21 | 0.13 |
| 500 | 1.53 | 3.07 | 1.49 | 1.08 | 0.37 | 3.18 | 1.49 | 1.08 | 0.36 | 2.35 | 0.58 | 0.40 | 0.24 |
| 1000 | 14.55 | **25.24** | 12.25 | 9.06 | 5.20 | 22.56 | 13.11 | 7.2 | 4.37 | 18.22 | 6.31 | 4.49 | 3.10 |

### 3.2.5   Communication and Mapping

We use a sequential distribution of work where incremental computing nodes receive contiguous segments of the data. For example, with horizontal and vertical partitioning, each node $p_i$ will receive $\lceil n/p \rceil \times m$ and $n \times \lceil m/p \rceil$ records, respectively.

As $p_i$ constructs its $GST_i$, it stores it in the file system for further processing. Upon completion, all GSTs are sent via network to the nearest processor based on latency. For example, in Fig. 5, P3 and P4 will send their GST to P1 and P2 respectively and P1, P2 will merge these trees in parallel. We utilize external memory as the suffix tree are arbitrarily large for a genomic dataset and can overflow the main memory of a single computing node.

## 4   Experimental Results and Analysis

### 4.1   Evaluation Datasets and Implementation

We evaluate our framework on uniformly distributed synthetic datasets as it allows us to perturb the dimensions and check the performance of the underlying methods. Hence, we generate different datasets with $n, m \in \{200, 300, 400, 500, 1000\}$ and name them accordingly. Notably, the sequential algorithm could not finish for larger dataset ($D_{10,000}$) due to its memory requirement. We did not consider $n, m$ in millions due to our computational restrictions as we were only able to access a small computing cluster [1]. Our implementations along with the data are available in https://github.com/mominbuet/ParallelGST.

### 4.2   Performance Analysis

We analyze our proposed approach in terms of $n, m, p$ and all three (distributed, shared and hybrid) memory models. Here, the distributed memory model will

**Table 2.** Execution time (in seconds) of bi-directional partitioning to build GST on different datasets with number of processors $p = \{1, 4, 8, 16\}$

| Data | Serial | Distributed | | | Shared | | | Hybrid | | | |
|---|---|---|---|---|---|---|---|---|---|---|---|
| | 1 | 4 | 8 | 16 | 4 | 8 | 16 | 4 | 8 | 16 | 32 |
| 200 | 4.8 | 94.2 | 90 | 87 | 43.8 | 42.6 | 38.4 | 70.8 | 73.2 | 75.6 | 1.51 |
| 300 | 16.2 | 121.8 | 107.4 | 106.2 | 72 | 48.6 | 43.2 | 88.8 | 75 | 75.6 | 1.37 |
| 400 | 35.4 | 168.6 | 148.2 | 124.8 | 102.6 | 54 | 57.6 | 114 | 87 | 96 | 1.36 |
| 500 | 91.8 | 231.6 | 151.8 | 154.8 | 145.2 | 76.2 | 62.4 | 146.4 | 103.8 | 105 | 1.36 |
| 1000 | 873 | 1135.2 | 428.4 | 291.6 | 856.8 | 202.2 | 154.8 | 635.4 | 312 | 214.2 | 1.36 |
| *10,000* | – | 3191.7 | 428.5 | 79.6 | – | – | – | 2878 | 418 | 77.3 | 18.78 |

not incorporate any intra-node parallelism instructions as discussed in Sect. 3.2.3 whereas the hybrid method will utilize both.

The shared memory architecture distributes the work into different co-located processors (cores) on one single node. Notably, in this model, we do not require any communication between two processes whereas the distributed model will incur communicating the *GST*s. However, the number of processors and memory available in shared model is fixed and limited as we can add new machines in the distributed model. Nevertheless, this comparison will denote the difference in the two memory architecture.

In Tables 1 and 2 we show the execution time of horizontal, vertical and bi-directional partitioning, respectively. Each method is executed on $p = \{2, 4, 8, 16\}$ processors whereas $p = 1$ denotes the serial or sequential execution. The sequential method is plaintext Ukkonen's algorithm [14]. Furthermore, the proposed hybrid approach uses both distributed and shared memory model with two cores on each processor of distributed machines for the 0 and 1 branches of GSTs.

In Table 1, The GST building time for smaller datasets ($n, m \leq 200$) are almost same for all settings. However, as the dataset size increases, the difference in execution time starts to diverge. For example, the sequential execution of $D_{200}$ takes 0.08 min whereas $D_{1000}$ requires 14.55 min. The same operation takes 3.08 min on the hybrid approach with $p = 16$. Similarly, the distributed model takes 6.09 min which shows the impact of intra-node parallelism.

However, one interesting outcome is the shared model's performance. It takes the minimum time of 2.28 min with $p = 16$ which is the lowest in all three experimental settings. However, it is noteworthy that it ran out of memory for datasets $n, m > 1000$. This depicts the necessity of the distributed or hybrid model as shared memory model are more suitable for datasets which only fits the main memory.

Table 1 also shows the impact of vertical partitioning where we need to add the path graphs. This addition is the only difference from the horizontal approach as all the nodes working on data $m_i < m$, needs to merge $m - m_i$ characters to the underlying GSTs. For example, with vertical method it takes 25.24 min

**Table 3.** Maximum Execution time (seconds) of Tree Building (TB), Add Path (AP) and Tree Merge (TM) for $D_{1000}$

| $p$ | Horizontal | | | Vertical | | | Bi-directional | | |
|---|---|---|---|---|---|---|---|---|---|
| | TB | AP | TM | TB | AP | TM | TB | AP | TM |
| 4 | 113.35 | – | 70.02 | 292.97 | 2.7 | 66.8 | 4.01 | 0.37 | 3.85 |
| 8 | 47.38 | – | 85.4 | 138.87 | 2.9 | 61.1 | 0.62 | 0.16 | 1.8 |
| 16 | 15.6 | – | 98 | 64.4 | 3.2 | 57.6 | 0.12 | 0.07 | 1.2 |

**Table 4.** Speedup analysis on $D_{1000}$ for all methods with $p = \{2, 4, 8, 16\}$

| Method | Distributed | | | | Shared | | | | Hybrid | | | |
|---|---|---|---|---|---|---|---|---|---|---|---|---|
| | 2 | 4 | 8 | 16 | 2 | 4 | 8 | 16 | 2 | 4 | 8 | 16 |
| Horizontal | 0.58 | 1.19 | 1.61 | 2.80 | 0.64 | 1.11 | 2.02 | 3.33 | 0.80 | 2.31 | 3.24 | 4.69 |
| Vertical | 0.90 | 1.74 | 2.31 | 2.39 | 0.84 | 2.62 | 4.45 | 6.38 | 0.85 | 3.48 | 4.66 | 4.72 |
| Bi-directional | – | 0.77 | 2.04 | 2.99 | – | 1.02 | 4.32 | 5.64 | – | 1.37 | 2.80 | **4.08** |

to process $D_{1000}$ whereas it took only 16.23 on horizontal approach. The rest of the execution time also follows the same trend as more processor leads to faster executions overall. The performance gain with shared model compared to hybrid is also lost due to the thread synchronization as the threads operate on $m_i < m$ requires more time for sequential path graph addition.

In Table 2, we show our best results where the data is partitioned into both directions. Here, the tree building cost is reduced compared to the prior two approaches as it resulted in smaller sub-trees. For example, with $n = m = 100$ and $p = 4$, each processor $p_i$ will work on $25 \times 25$ sized matrix whereas it will lead to $25 \times 100$ and $100 \times 25$ partitions for horizontal and vertical, respectively.

Table 3 demonstrates the granular execution time for tree building, path graph addition and the merge operation. We took the maximum time from each run as these functions were executed in parallel. Notably, these values are the building blocks for Tables 1 and 2. For example, the tree building time decreases with the increment of processors $p$. Furthermore, the bi-directional tree build cost decrements with the increment in processors as it divides the data by half.

In Table 4 we summarize the speedup $(= T_{par}/T_{seq})$ results for $D_{1000}$. Here, the shared model performs well compared to distributed model due to its zero communication cost. Notably, the distributed one is competitive for all $p > 2$ cases. Nevertheless, the shared model could not finish the $D_{10,000}$ as it ran out of the shared memory. On the contrary, both distributed and our hybrid model constructed the targeted GST as it did not depend on the limited, fixed main memory of one machine.

### 4.3   Utility Measure

We show the execution time of the targeted queries in Table 5. It denotes the efficiency of GSTs performing arbitrary string search as the time only increases

**Table 5.** Query 1, 2 and 3 execution time in seconds

| Query length $|q|$ | $D_{1000}$ | | | $D_{500}$ | | |
|---|---|---|---|---|---|---|
| | EM | SMM | PVSMM | EM | SMM | PVSMM |
| 400 | 5e4 | 0.14 | 0.15 | 4e4 | 0.13 | 0.12 |
| 500 | 6e4 | 0.21 | 0.22 | 5e4 | 0.19 | 0.18 |
| 1000 | 1e3 | 0.68 | 0.72 | 1e3 | 0.63 | 0.59 |

**Table 6.** Design-level comparison of previous works and ours in GST construction

| Work | Parallelism model | | Disk-based | GST |
|---|---|---|---|---|
| | Distributed | Shared | | |
| TDD [13] | ✗ | ✗ | ✓ | ✗ |
| TRELLIS [10] | ✗ | ✗ | ✓ | ✗ |
| Wavefront [6] | ✓ | ✗ | ✓ | ✗ |
| $ER_A$ [8] | ✓ | ✗ | ✓ | ✗ |
| PCF [4] | ✓ | ✗ | ✗ | ✗ |
| Shun and Blelloch [11] | ✗ | ✓ | ✗ | ✗ |
| DGST [15] | ✓ | ✗ | ✓ | ✓ |
| Our work | ✓ | ✓ | ✓ | ✓ |

with the query length $|q|$. Notably, the execution time varies slightly for different datasets as we only show the time for $D_{1000}$ and $D_{500}$ for space limitations.

## 5  Related Works

There has been multiple attempts in our targeted problem as shown in Table 6. Since GST of a large genomic dataset does not fit a sizeable memory, there have been several works to construct the tree in a file system [5]. These disk-based suffix trees usually store the individual subtree (s) on file similar to our approach [12]. For example, Tian *et al.* [13] showed a different suffix tree merging method *ST-Merge* using the Top-Down Disk (TDD) Algorithm.

Wavefront [6] and its successor $ER_A$ (Elastic Range) [8] both targeted disk-based and parallel approach to construct suffix trees. However, these works only considered a suffix tree and distributed memory model, whereas, in this work, we propose a hybrid method and GST. Comin and Farreras [4] proposed Parallel Continuous Flow (PCF) which efficiently distributes the lexical sorting process into multiple processors. Analogous to this work, Shun and Blelloch [11] also proposed a parallel construction scheme utilizing *cilk* (shared memory) in 2014. However, both works target suffix trees whereas GSTs contain a large number of sequences which is more complicated and at the same time more useful.

Finally, in a very recent work in 2019, DGST [15] offered a 3× speed up with data-parallel platform Spark and performed better than the state-of-the-art $ER_A$ [8]. Nevertheless, it did not employ the shared or hybrid model as we performed better with 4× speedup. One work in 2016 did report speedup upto 6× utilizing parallelism from Graphics Processing Units [9]. However, we do not use such H/W and could not benchmark as their implementations is unavailable.

# 6   Conclusion

In this paper, we constructed GSTs for genomic data in parallel using external memory. We also analyzed its performance using different datasets and queries. In future, we would like to investigate parallel and private query execution on the suffix/prefix tree structure [3]. Moreover, our methods can be utilized for constructing suffix arrays and benchmarked accordingly. Nevertheless, the proposed parallel constructions can be generalized for other tree-based data structures (*e.g.*, prefix) which can be useful for different genomic data computations.

**Acknowledgments.** The research is supported in part by the CS UManitoba Computing Clusters, Amazon Research Grant and NSERC Discovery Grants (RGPIN-2015-04147).

# References

1. Computing Resources. www.cs.umanitoba.ca/computing. Accessed 4 Dec 2019
2. Bieganski, P., Riedl, J., Carlis, J.V., Retzel, E.F.: Generalized suffix trees for biological sequence data: applications and implementation. In: HICSS (5), pp. 35–44 (1994)
3. Chen, L., Aziz, M.M., Mohammed, N., Jiang, X.: Secure large-scale genome data storage and query. Comput. Methods Programs Biomed. **165**, 129–137 (2018)
4. Comin, M., Farreras, M.: Parallel continuous flow: a parallel suffix tree construction tool for whole genomes. J. Comput. Biol. **21**(4), 330–344 (2014)
5. Farach, M., Ferragina, P., Muthukrishnan, S.: Overcoming the memory bottleneck in suffix tree construction. In: Proceedings 39th FOCS, pp. 174–183. IEEE (1998)
6. Ghoting, A., Makarychev, K.: Serial and parallel methods for i/o efficient suffix tree construction. In: Proceedings of the 2009 ACM SIGMOD International Conference on Management of Data, pp. 827–840. ACM (2009)
7. Hariharan, R.: Optimal parallel suffix tree construction. J. Comput. Syst. Sci. **55**(1), 44–69 (1997)
8. Mansour, E., Allam, A., Skiadopoulos, S., Kalnis, P.: ERA: efficient serial and parallel suffix tree construction for very long strings. Proc. VLDB **5**(1), 49–60 (2011)
9. Mišić, M.J., et al.: Parallelization of GST algorithm for source code similarity detection. In: 24th TELFOR, pp. 1–4. IEEE (2016)
10. Phoophakdee, B., Zaki, M.J.: Genome-scale disk-based suffix tree indexing. In: SIGMOD International Conference on Management of Data, pp. 833–844. ACM (2007)

11. Shun, J., Blelloch, G.E.: A simple parallel cartesian tree algorithm and its application to parallel suffix tree construction. ACM TOPC **1**(1), 8 (2014)
12. Tata, S., Hankins, R.A., Patel, J.M.: Practical suffix tree construction. In: Proceedings of the 13th International Conference VLDB, pp. 36–47 (2004)
13. Tian, Y., Tata, S., Hankins, R.A., Patel, J.M.: Practical methods for constructing suffix trees. VLDB J. **14**(3), 281–299 (2005)
14. Ukkonen, E.: Online construction of suffix trees. Algorithmica **14**(3), 249–260 (1995)
15. Zhu, G., et al.: DGST: efficient and scalable suffix tree construction on distributed data-parallel platforms. Parallel Comput. **87**, 87–102 (2019)

# A 3.5-Approximation Algorithm for Sorting by Intergenic Transpositions

Andre Rodrigues Oliveira[1]([✉]) [ID], Géraldine Jean[2] [ID], Guillaume Fertin[2] [ID],
Klairton Lima Brito[1] [ID], Ulisses Dias[3] [ID], and Zanoni Dias[1] [ID]

[1] Institute of Computing, University of Campinas, Campinas, Brazil
{andrero,klairton,zanoni}@ic.unicamp.br
[2] LS2N, UMR CNRS 6004, University of Nantes, Nantes, France
{geraldine.jean,guillaume.fertin}@univ-nantes.fr
[3] School of Technology, University of Campinas, Limeira, Brazil
ulisses@ft.unicamp.br

**Abstract.** Genome Rearrangements affect large stretches of genomes during evolution. One of the most studied genome rearrangement is the *transposition*, which occurs when a sequence of genes is moved to another position inside the genome. Mathematical models have been used to estimate the evolutionary distance between two different genomes based on genome rearrangements. However, many of these models have focused only on the (order of the) genes of a genome, disregarding other important elements in it. Recently, researchers have shown that considering existing regions between each pair of genes, called *intergenic regions*, can enhance the distance estimation in realistic data. In this work, we study the transposition distance between two genomes, but we also consider intergenic regions, a problem we name Sorting Permutations by Intergenic Transpositions (SbIT). We show that this problem is NP-hard and propose a 3.5-approximation algorithm for it.

**Keywords:** Genome rearrangements · Intergenic regions ·
Transpositions · Approximation algorithm

## 1 Introduction

Genome rearrangements are events that modify genomes by inserting or removing large stretches of DNA sequences, or by changing the order and the orientation of genes inside genomes. A transposition [1] is a rearrangement that swaps the position of two adjacent sequences of genes inside a genome. Another example of genome rearrangement is the reversal [11], that reverses the order and the orientation of a sequence of genes.

We compute the **rearrangement distance** between two genomes by determining the minimum number of events that transform one into another. A **model** $\mathcal{M}$ is a set of genome rearrangements that can be used to calculate the rearrangement distance.

© Springer Nature Switzerland AG 2020
C. Martín-Vide et al. (Eds.): AlCoB 2020, LNBI 12099, pp. 16–28, 2020.
https://doi.org/10.1007/978-3-030-42266-0_2

Algorithms based on the rearrangement distance perform whole-genome comparison and may be used as a tool to infer phylogenetic relationships. The usual method fills a matrix of pairwise distances among genomes that is later used to generate phylogenetic trees [2,12,14]. As for "classical" rearrangements, having a large spectrum of models globally helps better-understanding things.

While in practice it is likely rarely so, if genomes contain no repeated gene and share the same set of $n$ genes, they can be represented as permutations. Without loss of generality, we consider that one of these genomes is the **identity permutation**, i.e., the sorted permutation $\iota = (1\ 2\ \ldots n)$.

The Sorting by Rearrangements Problem thus consists in determining the shortest sequence of events from $\mathcal{M}$ that sorts a permutation $\pi$, i.e., that transforms it to $\iota$. Sorting by Rearrangements has been extensively studied in the past. For instance, **Sorting by Transpositions** has been proved NP-hard [6], while the best algorithm so far has an approximation factor of 1.375 [8].

Representing genomes through their gene order (thus, by permutations) implies that information not contained directly in the genes is lost. In particular, in the case of intergenic regions, DNA sequences between the genes are not considered. Recently, some authors argued that incorporating intergenic regions sizes in the models changes the distance estimations, and actually improves them [3,4]. It seems worth investigating models considering both gene order and intergenic sizes.

Results considering intergenic sizes for models with Double-Cut and Join (DCJ) and DCJs along with indels (i.e., insertions and deletions) are known: the former is NP-hard and it has a 4/3-approximation algorithm [9], while the latter is polynomial [7]. In addition to the approximation algorithm, the authors in [7] also developed two exact algorithms: a fixed-parameter tractable algorithm and an integer linear programming formulation. Practical tests from [7] showed that statistical properties of the inferred scenarios using intergenic regions are closer to the true ones than scenarios which do not use them.

Some results considering intergenic sizes with super short operations (i.e., a reversal or a transposition applied to one or two genes of the genome) are known [13]. Using the concept of breakpoints, Brito et al. showed a 4-approximation algorithm (resp. 6-approximation algorithm) for sorting by reversals (resp. reversals and transpositions) when also considering intergenic regions on unsigned permutations [5]. They also showed that both problems are NP-hard.

In this paper, we investigate the transposition distance between genomes that also takes into account intergenic regions, a problem we name **Sorting by Intergenic Transpositions (SbIT)**. Instead of using breakpoints, here we propose a modification of a known graph structure to represent both gene order and intergenic sizes in a single graph. We show that SbIT is NP-hard, and, with the help of this adapted graph structure, we design a 3.5-approximation algorithm.

This work is organized as follows. Section 2 presents some definitions we extensively use throughout this paper. Section 3 presents the graph structure we use to produce our approximation algorithm. Section 4 contains a series of intermediate lemmas that support our algorithm. Section 5 describes the 3.5-approximation algorithm for SbIT. Section 6 concludes the paper.

## 2   Basic Definitions

A genome $\mathcal{G}$ is a sequence of $n$ genes denoted by $g_i$, with $i \in [1..n]$, in which two consecutive genes $g_{j-1}$ and $g_j$, with $j \in [2..n]$, are separated by a noncoding region called intergenic region, denoted by $r_j$ – that are also present on its extremities ($r_1$ and $r_{n+1}$): $\mathcal{G} = r_1, g_1, r_2, g_2, \ldots, r_n, g_n, r_{n+1}$.

Given two genomes of closely related species, for the simplifying purposes of our initial analysis, we expect they will share the same set of genes, which may appear in different orders due to genome rearrangements. Selective pressures tend to conserve genes and not intergenic regions [3]. Therefore, genome rearrangements hardly cut inside genes, whereas cuts appear in intergenic regions.

Our model assumes that (i) no gene is duplicated in a genome and (ii) both genomes share the same set of genes. We assign unique integer numbers in the range $[1..n]$ to each gene and represent them as a permutation. Therefore, the sequence of genes in a genome is modeled by a permutation $\pi = (\pi_1 \ \pi_2 \ \ldots \ \pi_n)$, $\pi_i \in \mathbb{N}$, $1 \le \pi_i \le n$, and $\pi_i \ne \pi_j$ for all $i \ne j$.

We represent intergenic regions by their lengths instead of assigning unique identifiers to each of them, which would be pointless because rearrangements may split intergenic regions several times. The sequence of intergenic regions around $n$ genes is represented as $\breve{\pi} = (\breve{\pi}_1 \ \breve{\pi}_2 \ \ldots \ \breve{\pi}_{n+1})$, $\breve{\pi}_i \in \mathbb{N}$. Intergenic region $\breve{\pi}_i$ is on the left side of $\pi_i$, whereas $\breve{\pi}_{i+1}$ is on the right side.

Our goal is to compute the distance between two genomes: $(\pi, \breve{\pi})$ and $(\sigma, \breve{\sigma})$. We may assign unique labels to genes arbitrarily, so we simplify the definition of our problem by setting $\sigma$ as the identity permutation $\iota$, such that $\sigma = \iota = (1 \ 2 \ \ldots \ n)$ and $\breve{\sigma} = \breve{\iota}$, which describes all the information we need to our problem. Therefore, an **instance** of our problem is composed by three elements $(\pi, \breve{\pi}, \breve{\iota})$, such that $\sum_{i=1}^{n+1} \breve{\pi}_i = \sum_{i=1}^{n+1} \breve{\iota}_i$, which guarantees that total intergenic region lengths are conserved.

An **intergenic transposition** is an operation $\rho_{(x,y,z)}^{(i,j,k)}$, $1 \le i < j < k \le n+1$, $0 \le x \le \breve{\pi}_i$, $0 \le y \le \breve{\pi}_j$, $0 \le z \le \breve{\pi}_k$, and $\{x, y, z\} \subset \mathbb{N}$. An intergenic transposition acts on instances to generate new ones: $(\pi, \breve{\pi}, \breve{\iota}) \cdot \rho_{(x,y,z)}^{(i,j,k)} = (\pi', \breve{\pi}', \breve{\iota})$, where (i) $\pi' = (\pi_1 \pi_2 \ldots \pi_{i-1} \ \pi_j \pi_{j+1} \ldots \pi_{k-1} \ \pi_i \pi_{i+1} \ldots \pi_{j-1} \ \pi_k \pi_{k+1} \ldots \pi_n)$, and (ii) $\breve{\pi}' = (\breve{\pi}_1 \ldots \breve{\pi}_{i-1} \ \boxed{\breve{\pi}'_i} \ \breve{\pi}_{j+1} \ldots \breve{\pi}_{k-1} \ \boxed{\breve{\pi}'_j} \ \breve{\pi}_{i+1} \ldots \breve{\pi}_{j-1} \ \boxed{\breve{\pi}'_k} \ \breve{\pi}_{k+1} \ldots \breve{\pi}_{n+1})$, such that $\breve{\pi}'_i = x + \breve{\pi}_j - y$, $\breve{\pi}'_j = z + \breve{\pi}_i - x$, and $\breve{\pi}'_k = y + \breve{\pi}_k - z$.

As we can see, while $\rho_{(x,y,z)}^{(i,j,k)}$ keeps $\breve{\iota}$ intact, it moves segments from $\pi$ and $\breve{\pi}$ to other positions and also modifies the contents of three elements from $\breve{\pi}$: it cuts $\breve{\pi}_i$ after first $x$ nucleotides, $\breve{\pi}_j$ after first $y$ nucleotides, and $\breve{\pi}_k$ after first $z$ nucleotides, and rearranges them as defined above. Figure 1 shows examples of instances and the application of an intergenic transposition.

The **intergenic transposition distance** $d_t(\pi, \breve{\pi}, \breve{\iota})$ is the minimum number $m$ of intergenic transpositions $\rho_1, \ldots, \rho_m$ that transform $\pi$ into $\iota$, and $\breve{\pi}$ into $\breve{\iota}$.

Therefore, $d_t(\pi, \breve{\pi}, \breve{\iota}) = m$ implies a minimal sequence $(\pi, \breve{\pi}, \breve{\iota}) \cdot \rho_1 \cdot \ldots \cdot \rho_m = (\iota, \breve{\iota}, \breve{\iota})$.

**Lemma 1.** *Sorting by Intergenic Transpositions (SbIT) is NP-hard.*

*Proof.* The Sorting by Transpositions problem (SbT) has already been proved NP-hard [6]. An instance of this problem consists of a permutation $\gamma$ and a non-negative integer $d$. The goal is to determine if its possible to transform $\gamma$ into $\iota$ applying at most $d$ transpositions.

We can reduce all instances of SbT to instances of SbIT by setting $\pi = \gamma$ and $\breve{\pi} = \breve{\iota} = (0\ 0\ \ldots\ 0)$. Note that it is possible to transform $\gamma$ into $\iota$ applying at most $d$ transpositions if and only if $d_t(\pi, \breve{\pi}, \breve{\iota}) \leq d$.     $\square$

From now on, we will refer to intergenic transposition as transposition only.

## 3   Weighted Cycle Graph

We adapted a graph structure called *breakpoint graph* [1,10] to conveniently represent an instance $(\pi, \breve{\pi}, \breve{\iota})$ in a single graph. This structure allows us to describe algorithms and prove approximation bounds. All definitions we propose here are exemplified in Figs. 2 and 3.

We represent a given instance by a **weighted cycle graph** $G(\pi, \breve{\pi}, \breve{\iota}) = (V, E, w)$, where $V$ is the set $\{-n, \ldots, -2, -1, 1, 2, \ldots, n\} \cup \{0, -(n+1)\}$, $E$ is the set of edges that can be either gray or black, and $w : E \to \mathbb{N}$ is a function mapping edges to values corresponding to intergenic region lengths. The black edge set is $\{e_i = (-\pi_i, +\pi_{i-1}) : 1 \leq i \leq n+1\}$, and $w(e_i) = \breve{\pi}_i$. The gray edge set is $\{e'_i = (+(i-1), -i) : 1 \leq i \leq n+1\}$, and $w(e'_i) = \breve{\iota}_i$. In this definition, we consider $\pi_0 = 0$ and $\pi_{n+1} = n+1$.

The graph can be drawn in many arbitrary ways, but it is more convenient to place its vertices on a horizontal line in the same order as the elements of $\pi$, so $\pi_0$ (resp. $-\pi_{n+1}$) is the leftmost (resp. rightmost) element of it. In addition,

$\mathcal{G}_1$ : CAT ⟦Gene 1⟧ AACCG ⟦Gene 2⟧ T ⟦Gene 3⟧ CTGTA ⟦Gene 4⟧ ACTCAC ⟦Gene 5⟧ GGC ⟦Gene 6⟧ TGTTCTG ⟦Gene 7⟧ GA ⟦Gene 8⟧ CT  $(\iota)$

$\mathcal{G}_2$ : G ⟦Gene 3⟧ TTCAGAT ⟦Gene 2⟧ ⟦Gene 1⟧ CTCT ⟦Gene 7⟧ CT ⟦Gene 6⟧ AGTGACA ⟦Gene 4⟧ ATCC ⟦Gene 8⟧ ⟦Gene 5⟧ TGGCTCTGA  $(\pi)$

$\rho_{(2,1,2)}^{(2,5,6)}$

$\mathcal{G}_3$ : G ⟦Gene 3⟧ TTT ⟦Gene 6⟧ AGCAGAT ⟦Gene 2⟧ ⟦Gene 1⟧ CTCT ⟦Gene 7⟧ CTGACA ⟦Gene 4⟧ ATCC ⟦Gene 8⟧ ⟦Gene 5⟧ TGGCTCTGA  $(\pi')$

**Fig. 1.** Two (fictitious) genomes $\mathcal{G}_1$ and $\mathcal{G}_2$ that share 8 genes. We represent $\mathcal{G}_1$ as the identity permutation, which leads to $\mathcal{G}_2$ as the permutation $\pi = (3\ 2\ 1\ 7\ 6\ 4\ 8\ 5)$. We assume that the number of nucleotides between genes are good estimators for intergenic regions lengths. For example, in $\mathcal{G}_1$ before "Gene 1" we have 3 nucleotides, between "Gene 1" and "Gene 2" we have 5, and so on. Thus, $(\pi, \breve{\pi}, \breve{\iota})$ is such that $\pi = (3\ 2\ 1\ 7\ 6\ 4\ 8\ 5)$, $\breve{\pi} = (1\ 7\ 0\ 4\ 2\ 7\ 4\ 0\ 9)$, and $\breve{\iota} = (3\ 5\ 1\ 5\ 6\ 3\ 7\ 2\ 2)$. Genome $\mathcal{G}_3$ represents $(\pi', \breve{\pi}', \breve{\iota}') = (\pi, \breve{\pi}, \breve{\iota}) \cdot \rho_{(2,1,2)}^{(2,5,6)}$, so $\pi' = (3\ 6\ 2\ 1\ 7\ 4\ 8\ 5)$, $\breve{\pi}' = (1\ 3\ 7\ 0\ 4\ 6\ 4\ 0\ 9)$.

for each element $\pi_i \in \pi$, vertex $-\pi_i \in G(\pi, \breve{\pi}, \breve{\iota})$ is drawn to the left of vertex $+\pi_i$. Since black edges relate to $\pi$, they are drawn as horizontal lines, and we label the black edge $e_i$ as $i$. Gray edges are drawn as arcs.

Each vertex in $G(\pi, \breve{\pi}, \breve{\iota})$ has a gray edge and a black edge, which allows a unique decomposition of edges in cycles of alternating colors. Each cycle $C$ with $\ell$ black edges is represented as a list $(c^1, c^2, \ldots, c^\ell)$ of the labels from its black edges, and to make the notation unique we assume $c^1$ to be the index of the "rightmost" black edge (i.e., the black edge with the highest label using our default drawing) and we traverse it from right to left. We follow by several definitions regarding cycles.

A cycle is **long** if it has 3 or more black edges; a cycle is **short** if it has 2 black edges; a cycle is **trivial** if it has 1 black edge; a cycle is **non-trivial** if it is either short or long. A non-trivial cycle $C = (c^1, \ldots, c^\ell)$ is **non-oriented** if $c^1, \ldots, c^\ell$ is a decreasing sequence (note that every short cycle $C$ is non-oriented); $C$ is **oriented** otherwise.

Given a non-trivial cycle $C = (c^1, \ldots, c^\ell)$, every pair of black edges $e_{c^i}$ and $e_{c^{(i+1)}}$ with $1 \le i < \ell$ is called an **open gate** if for any $c^j \in C$ with $j \notin \{i, i+1\}$ either $c^j > c^i$ or $c^j < c^{i+1}$, if $c^i > c^{i+1}$, and either $c^j > c^{i+1}$ or $c^j < c^i$ otherwise. Besides, the pair of black edges $e_{c^1}$ and $e_{c^\ell}$ is called an **open gate** if $c^1 > c^j > c^\ell$ for any $c^j \in C$ with $j \notin \{1, \ell\}$. Note that on every short cycle $C = (c^1, c^2)$ the pair $e_{c^1}, e_{c^2}$ is an open gate.

Bafna and Pevzner [1] showed that for every open gate $e_{c^i}$ and $e_{c^j}$ from $C$ with $c^i > c^j$ there exists another non-trivial cycle $D$ with black edges $e_{d^i}$ and $e_{d^j}$ such that either $c^i > d^i > c^j > d^j$ or $d^i > c^i > d^j > c^j$. In this case, we say that $D$ **closes** this open gate. Figure 2 shows two examples of cycles with open gates and how cycles can close these open gates.

Cycles can be either **balanced** or **unbalanced**. A cycle $C = (c^1, \ldots, c^\ell)$ is **balanced** if $\sum_{i=1}^\ell |w(e'_{c^i}) - w(e_{c^i})| = 0$, and is **unbalanced** otherwise. In other words, one cycle is balanced if the sum of weights of gray edges equals the sum of weights of black edges, and it is unbalanced otherwise. An unbalanced cycle $C = (c^1, \ldots, c^\ell)$ is called **positive** if $\sum_{i=1}^\ell (w(e'_{c^i}) - w(e_{c^i})) > 0$, and it is called **negative** otherwise. Figure 3 shows an example of a weighted cycle graph and

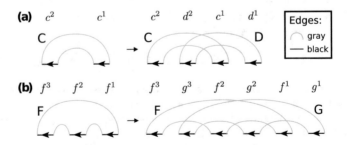

**Fig. 2.** (a) The cycle $C = (c^1, c^2)$ has an open gate $(c^1, c^2)$ closed by $D = (d^1, d^2)$ on the right. (b) The cycle $F = (f^1, f^2, f^3)$ has three open gates $(f^1, f^2)$, $(f^2, f^3)$, and $(f^3, f^1)$ closed by $G = (g^1, g^2, g^3)$ on the right.

the application of an intergenic transposition on it, as well as the weighted cycle graph for an instance $(\iota, \breve{\iota}, \breve{\iota})$.

Let $c(\pi, \breve{\pi}, \breve{\iota})$, $c_b(\pi, \breve{\pi}, \breve{\iota})$, and $c_u(\pi, \breve{\pi}, \breve{\iota})$, denote the number of cycles, balanced cycles, and unbalanced cycles in $G(\pi, \breve{\pi}, \breve{\iota})$, respectively.

**Lemma 2.** *The instance $(\iota, \breve{\iota}, \breve{\iota})$ has two properties that do not occur together in any other instance: (i) $c(\iota, \breve{\iota}, \breve{\iota}) = n + 1$, and (ii) $c_b(\iota, \breve{\iota}, \breve{\iota}) = n + 1$. As a consequence, we have that $c_u(\iota, \breve{\iota}, \breve{\iota}) = 0$ (see Fig. 3(c) for an example).*

Note that Sorting by Intergenic Transpositions is more complicated than the Sorting by Transpositions because increasing the number of cycles is not sufficient - these cycles need to be balanced.

Given a sequence of transpositions $\mathcal{S}_\rho = (\rho_1, \rho_2, \ldots, \rho_k)$, let $(\pi, \breve{\pi}, \breve{\iota}) \cdot \mathcal{S}_\rho$ denotes $(\pi, \breve{\pi}, \breve{\iota}) \cdot \rho_1 \cdot \rho_2 \cdot \ldots \cdot \rho_k$ such that $\rho_{i+1}$ is always a transposition for $(\pi, \breve{\pi}, \breve{\iota}) \cdot \rho_1 \cdot \ldots \cdot \rho_i$ with $1 \le i < k$. Let $\Delta c(\pi, \breve{\pi}, \breve{\iota}, \mathcal{S}_\rho) = c((\pi, \breve{\pi}, \breve{\iota}) \cdot \mathcal{S}_\rho) - c(\pi, \breve{\pi}, \breve{\iota})$ and $\Delta c_b(\pi, \breve{\pi}, \breve{\iota}, \mathcal{S}_\rho) = c_b((\pi, \breve{\pi}, \breve{\iota}) \cdot \mathcal{S}_\rho) - c_b(\pi, \breve{\pi}, \breve{\iota})$ denote the variation in the number of cycles and balanced cycles, respectively, when $\mathcal{S}_\rho$ is applied to $(\pi, \breve{\pi}, \breve{\iota})$.

**Lemma 3.** $\Delta c(\pi, \breve{\pi}, \breve{\iota}, \rho) \in \{2, 0, -2\}$ *for any transposition $\rho$.*

*Proof.* Straightforward from Bafna and Pevzner [1]. □

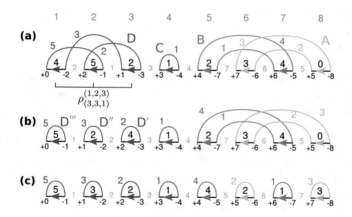

**Fig. 3.** We color the figure to improve cycle visualization. Black edges are drawn as horizontal lines, and gray edges are drawn as arcs. Arrows over black edges represent how we traverse them, and numbers on the top of the image indicate black edges labels. **(a)** The weighted cycle graph $G(\pi, \breve{\pi}, \breve{\iota})$ for $\pi = (2\ 1\ 3\ 4\ 7\ 6\ 5)$, $\breve{\pi} = (4\ 5\ 2\ 1\ 2\ 3\ 4\ 0)$, and $\breve{\iota} = (5\ 3\ 2\ 1\ 4\ 2\ 1\ 3)$. $G(\pi, \breve{\pi}, \breve{\iota})$ has four cycles: $A = (8, 6)$ is a non-oriented positive short cycle; $B = (7, 5)$ is a non-oriented short negative cycle; $C = (4)$ is a trivial balanced cycle, and $D = (3, 1, 2)$ is an oriented long negative cycle. **(b)** The weighted cycle graph $G((\pi, \breve{\pi}, \breve{\iota}) \cdot \rho^{(1,2,3)}_{(3,3,1)})$, with $\Delta c(\pi, \breve{\pi}, \breve{\iota}, \rho) = 2$ and $\Delta c_b(\pi, \breve{\pi}, \breve{\iota}, \rho) = 1$. The transposition applied to the negative cycle $D$ from **(a)** transformed it into three trivial cycles $D' = (3)$, $D'' = (2)$, and $D''' = (1)$ such that $D'''$ is balanced. **(c)** The weighted cycle graph $G(\iota, \breve{\iota}, \breve{\iota})$, with $c(\iota, \breve{\iota}, \breve{\iota}) = c_b(\iota, \breve{\iota}, \breve{\iota}) = n + 1$, and $c_u(\iota, \breve{\iota}, \breve{\iota}) = 0$.

**Lemma 4.** $\Delta c_b(\pi, \check{\pi}, \check{\iota}, \rho) \leq 2$ *for any transposition* $\rho$.

*Proof.* From Lemma 3 we know that we can increase the number of cycles by at most 2. In this scenario, one cycle $C$ is split in three by a single transposition $\rho$. If $C$ is balanced, the best we can expect is that $\rho$ creates three balanced cycles, so $\Delta c_b(\pi, \check{\pi}, \check{\iota}, \rho) = 2$. Otherwise, at least one of the resulting cycles shall be unbalanced too, since weights of black edges of the three cycles sum up to a value that is different from the sum of weights of gray edges. Therefore, the best we expect is that the other two cycles are balanced, so $\Delta c_b(\pi, \check{\pi}, \check{\iota}, \rho) = 2$.    □

**Theorem 5.** $d_t(\pi, \check{\pi}, \check{\iota}) \geq \frac{n+1-c_b(\pi, \check{\pi}, \check{\iota})}{2}$.

*Proof.* By Lemma 2 we know that $c_b(\iota, \check{\iota}, \check{\iota}) = n + 1$. Therefore, our goal is to increase the number of cycles from $c_b(\pi, \check{\pi}, \check{\iota})$ to $n+1$; since this number increases by at most 2 for each transposition (Lemma 4), the lemma follows.    □

## 4    Preliminary Results

This section presents properties and lemmas to support the 3.5-approximation presented in Sect. 5.

Let $e_{c^x}$ and $e_{c^y}$ be two arbitrary black edges in the same cycle $C = (c^1, \ldots, c^\ell)$, with $\{x, y\} \subset [1..\ell]$. We define the function $f : E \times E \to \mathbb{Z}$ as $f(e_{c^x}, e_{c^y}) = \sum_{i=x \ (\text{mod } \ell)+1}^{y-1} [w(e'_{c^i}) - w(e_{c^i})] + w(e'_{c^x})$.

In other words, given the path $P$ of black and gray edges that goes from $e_{c^x}$ to $e_{c^y}$, $f(e_{c^x}, e_{c^y})$ computes the sum of weights of gray edges in $P$ minus the weights of black edges from $P$ – excluding the black edges $e_{c^x}$ and $e_{c^y}$. Observe that we traverse the first black edge in the cycle from right to left, which means that the path that goes from $e_{c^x}$ to $e_{c^y}$ is different from the path that goes from $e_{c^y}$ to $e_{c^x}$.

Note that $f(e_{c^x}, e_{c^y}) + f(e_{c^y}, e_{c^x}) - w(e_{c^x}) - w(e_{c^y})$ indeed computes the sum of weights of all gray edges minus the sum of weights of all black edges. Therefore, a cycle $C$ is balanced if $f(e_{c^x}, e_{c^y}) + f(e_{c^y}, e_{c^x}) - w(e_{c^x}) - w(e_{c^y}) = 0$, positive if $f(e_{c^x}, e_{c^y}) + f(e_{c^y}, e_{c^x}) - w(e_{c^x}) - w(e_{c^y}) > 0$, and negative if $f(e_{c^x}, e_{c^y}) + f(e_{c^y}, e_{c^x}) - w(e_{c^x}) - w(e_{c^y}) < 0$.

We follow by presenting several lemmas, that will later be combined to prove the correctness of our 3.5-approximation algorithm. Let us first present the ideas behind them. Due to space constraints, proofs of Lemmas 6–13 are omitted. However, it can be seen in Fig. 4 how (and in which case) each of these lemmas is applied, as explained in detail right after the lemmas.

The next two lemmas deal with non-trivial negative cycles. Lemmas 6 and 8 show that it is always possible to increase balanced cycles by applying one transposition on negative cycles that are non-oriented and oriented, respectively.

**Lemma 6.** *Let $C$ be a non-trivial non-oriented negative cycle. There is a transposition that increases the number of balanced cycles by one.*

**Lemma 7.** *Let* $C = (c^1, \ldots, c^k)$ *be a non-trivial oriented cycle. If* $C$ *is not positive, there is a triple* $(c^x, c^y, c^z)$ *with* $c^x > c^z > c^y$ *and* $1 \le x < y < z \le k$ *such that* $0 \le f(e_{c^x}, e_{c^y}) \le w(c^x) + w(c^y)$, *or* $0 \le f(e_{c^y}, e_{c^z}) \le w(c^y) + w(c^z)$, *or* $0 \le f(e_{c^z}, e_{c^x}) \le w(c^z) + w(c^x)$.

**Lemma 8.** *Let* $C$ *be an oriented long negative cycle. There is a transposition that increases the number of balanced cycles by 1 and the number of cycles by 2.*

Now let us explain how to deal with trivial negative cycles. We use Lemma 9 as an intermediary step to the correctness of Lemma 10, that shows how many transpositions are needed to transform trivial negative cycles into trivial balanced.

**Lemma 9.** *It is possible to make any redistribution of weights of three distinct black edges* $e_{b^1}$, $e_{b^2}$, *and* $e_{b^3}$ *using two transpositions.*

**Lemma 10.** *Let* $G$ *be a weighted cycle graph in which all negative cycles are trivial. Then, there are two transpositions that increase the number of balanced cycles by 2.*

The next three lemmas dealing with non-trivial balanced cycles. Lemmas 11, 12 and 13 show that it is possible to increase balanced cycles by applying transpositions on oriented long balanced cycles, non-oriented long balanced cycles, and short balanced cycles, respectively.

**Lemma 11.** *Let* $C$ *be an oriented long balanced cycle. Then it is possible to increase the number of balanced cycles by two after at most three transpositions.*

**Lemma 12.** *Let* $C$ *be a non-oriented long balanced cycle in a graph with no oriented cycles. It is possible to increase the number of balanced cycles by four after at most seven transpositions.*

**Lemma 13.** *Let* $G$ *be a graph with no long cycles such that all cycles are balanced. If* $G$ *has short cycles it is possible to increase the number of balanced cycles by two after two transpositions.*

In Fig. 4(a) the blue cycle $A = (6, 4, 2)$ is non-oriented and negative, so we can apply Lemma 6 using the positive cycle $B = (7, 5)$, and the intergenic transposition $\rho_{(5,4,1)}^{(2,6,7)}$ generates in Fig. 4(b) the trivial balanced cycle $C = (2)$, and the non-trivial cycle $D = (7, 5, 3, 6)$, that in this case is negative and oriented. We then can use Lemma 8 on $D$, and the transposition $\rho_{(2,1,0)}^{(3,6,7)}$ generates in Fig. 4(c) the trivial balanced cycle $E = (3)$, the trivial cycle $F = (7)$, and the short cycle $G = (6, 4)$. Note that $F$ is negative and $G$ is balanced. At this stage, there is no long cycle, and we can use Lemma 10 on the trivial negative cycle $F$. This lemma requires a positive cycle to interact with the trivial negative, and $H = (8)$ is the only positive cycle. Since $H$ is also trivial, we need to borrow a black edge of a balanced cycle to apply the transposition, so let us use the short balanced cycle $G$. Two consecutive transpositions on these black edges (see

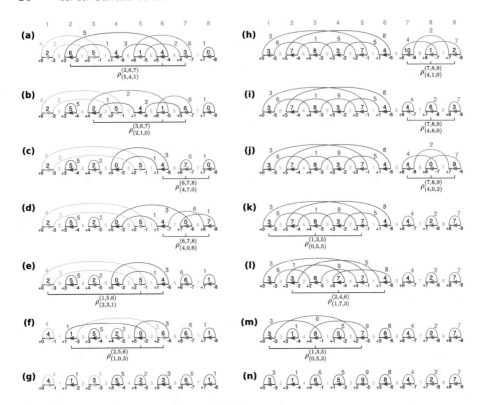

**Fig. 4.** Black egdes are drawn as horizontal lines, and their labels are on the top of the figure. Gray edges are drawn as arcs. **(a)–(g)** shows a sequence of intergenic transpositions that sorts the instance $(\pi, \breve{\pi}, \breve{\iota})$ with $\pi = (3\ 2\ 1\ 6\ 5\ 4\ 7)$, $\breve{\pi} = (2, 6, 5, 4, 1, 4, 3, 0)$, and $\breve{\iota} = (4, 1, 3, 5, 2, 3, 6, 1)$ using Lemmas 6–10 and 13. **(h)–(n)** shows a sequence of intergenic transpositions that sorts the instance $(\pi, \breve{\pi}, \breve{\iota})$ with $\pi = (5\ 4\ 3\ 2\ 1\ 6\ 8\ 7)$, $\breve{\pi} = (3, 7, 8, 3, 7, 4, 10, 1, 2)$, and $\breve{\iota} = (3, 1, 6, 5, 9, 8, 4, 2, 7)$ using Lemmas 11 and 12. (Color figure online)

Fig. 4(c) and (d)) generate two balanced cycles $I = (7)$ and $J = (8)$ in Fig. 4(e), without modifying $G$. The weighted cycle graph in Fig. 4(e) has only balanced cycles, and they are either short or trivial. In this case we can use Lemma 13 that applies two intergenic transpositions to two non-oriented short cycles, in this case $G = (6, 4)$ and $K = (5, 1)$. The two consecutive transpositions applied to $G$ and $K$ (see Fig. 4(e) and (f)) generate four balanced cycles, increasing the number of balanced cycles by 2, and completing the sorting process – the weighted cycle graph in Fig. 4(g) has only balanced cycles.

In Fig. 4(h) the green cycle $A' = (9, 7, 8)$ is oriented, and since it is also balanced we can use Lemma 11. The three transpositions in Fig. 4(h–j) breaks $A'$ into three balanced cycles. In Fig. 4(k) we have only balanced non-oriented cycles, and we can use Lemma 12 that applies a transposition followed by Lemma 11 twice. The first transposition is applied over cycle $B' = (5, 3, 1)$ transforming the blue cycle in Fig. 4(k) into an oriented balanced cycle $C' = (6, 2, 4)$. The first application of Lemma 11 breaks $C'$ into three balanced cycles (in this case, one transposition is sufficient) and also transforms the non-oriented cycle $B'$ into the oriented cycle $B'' = (5, 1, 3)$ (from Fig. 4(l) to (m)). The second application of Lemma 11 breaks $B''$ into three balanced cycles (using again only one transposition), completing the sorting process – Fig. 4(g) has only balanced cycles.

## 5    The 3.5-Approximation Algorithm

Algorithm 1 focuses on applying transpositions that increase the number of balanced cycles, following a sequence of steps using Lemmas 6, 8, 10, 11, 12, and 13.

---

**Algorithm 1.** a 3.5-approximation algorithm for SbIT.

---

**Data:** an instance $(\pi, \breve{\pi}, \breve{\iota})$.
**Result:** a sequence $\rho_1, \rho_2, \ldots, \rho_m$ such that $(\pi, \breve{\pi}) \cdot \rho_1 \cdot \rho_2 \cdot \ldots \cdot \rho_m = (\iota, \breve{\iota})$.

1   $sequence \leftarrow \emptyset$
2  **while** $(\pi, \breve{\pi}, \breve{\iota}) \neq (\iota, \breve{\iota}, \breve{\iota})$ **do**
3   | $G \leftarrow G(\pi, \breve{\pi}, \breve{\iota})$
4   | **if** there exists an oriented cycle $C$ in $G$ that is either balanced or negative **then**
5   | | **if** $C$ is negative **then**
6   | | | $S_\rho \leftarrow$ transposition from Lemma 8
7   | | **else** $S_\rho \leftarrow$ transpositions from Lemma 11
8   | **else if** there exists a negative cycle $C$ in $G$ that is either non-oriented or trivial **then**
9   | | **if** $C$ is non-oriented **then**
10  | | | $S_\rho \leftarrow$ transposition from Lemma 6
11  | | **else** $S_\rho \leftarrow$ transpositions from Lemma 10
12  | **else**
13  | | **if** there exists a long balanced cycle $C$ in $G$ **then**
14  | | | $S_\rho \leftarrow$ transpositions from Lemma 12
15  | | **else** $S_\rho \leftarrow$ transpositions from Lemma 13
16  | $(\pi, \breve{\pi}, \breve{\iota}) \leftarrow (\pi, \breve{\pi}, \breve{\iota}) \cdot S_\rho$
17  | $sequence.append(S_\rho)$
18 **return** $sequence$

---

Let us briefly show the correctness of Algorithm 1, i.e., it stops and reaches $(\iota, \breve{\iota}, \breve{\iota})$.

While $(\pi, \breve{\pi}, \breve{\iota}) \neq (\iota, \breve{\iota}, \breve{\iota})$ we have that one of the following must be true: (i) there is an oriented cycle. in this scenario we break this cycle if it is balanced or negative on lines 5 and 7; (ii) there is a negative cycle (considering that they are not oriented): we create balanced cycles if it is non-oriented or trivial on lines 9 and 11; and (iii) cycles are all balanced, and there is no oriented cycles. If there is a long cycle we break it at line 13, and we break short cycles at line 15.

Note that we did not care about the positive oriented cycles: they become either balanced or negative before the algorithm uses (iii), and will be handled in (i) at some point. If the algorithm reaches (iii) then all cycles are balanced since any negative cycle is handled by (i) and (ii).

Concerning the complexity of Algorithm 1, the loop of lines 2–17 iterates up to $m = n+1$ times. Since each time the algorithm applies one of those lemmas, it increases the number of balanced cycles by at least one. Finding which lemma to use (and at which positions the transposition takes place) requires $O(n^2)$ time. Thus, the overall complexity of Algorithm 1 is $O(n^3)$.

Now let us discuss about the approximation factor Algorithm 1 guarantees. Note that some of the steps of the algorithm require more than one transposition, so the approximation factor will be computed as follows:

**Definition 14.** *Let $S_\rho = (\rho_1, \rho_2, \ldots, \rho_d)$ be a sequence of transpositions such that $(\pi, \breve{\pi}, \breve{\iota}) \cdot S_\rho = (\sigma, \breve{\sigma}, \breve{\iota})$. By Lemma 4, $S_\rho$ creates up to 2d balanced cycles. Therefore, the approximation factor is at most $\frac{2d}{c_b(\sigma, \breve{\sigma}, \breve{\iota}) - c_b(\pi, \breve{\pi}, \breve{\iota})}$.*

The following lemma shows that each step of Algorithm 1 guarantees an approximation factor of 3.5 or less, which leads to the 3.5-approximation algorithm we propose.

**Lemma 15.** *Algorithm 1 has an approximation factor of 3.5.*

*Proof.* We use the formula from Definition 14 to calculate the approximation factor of each step.

- Step using Lemma 8 creates at least one new balanced cycle using one transposition, which leads to the maximum approximation $\frac{2}{1} = 2$.
- Step using Lemma 11 creates at least two new balanced cycles using up to three transpositions, so its maximum approximation factor is $\frac{6}{2} = 3$.
- Step using Lemma 6 creates a new balanced cycles using one transposition, and its approximation is $\frac{2}{1} = 2$.
- Step using Lemma 10 it creates two new balanced cycles using two transpositions, so its approximation is $\frac{4}{2} = 2$.
- Step using Lemma 12 creates four new balanced cycle using up to seven transpositions, and it follows that the maximum approximation is $\frac{14}{4} = 3.5$.
- Step using Lemma 13 creates two new balanced cycles using two transpositions, so it follows that its approximation is $\frac{4}{2} = 2$. □

# 6    Conclusion

We adapted the breakpoint graph to represent both gene order and intergenic sizes, and investigated properties of this new graph structure during a sorting process. As a result, we were able to design an approximation algorithm for the Sorting by Intergenic Transpositions. We also show that this problem is NP-Hard.

As future works, one can explore a problem where the probability of an intergenic region being affected by transpositions is related to its size, i.e., when genome rearrangements are more likely to cut the genome on bigger intergenic regions. One can also investigate the use of reversals and transpositions on signed permutations along with intergenic regions.

**Acknowledgments.** This work was supported by the National Council for Scientific and Technological Development - CNPq (grants 400487/2016-0, 425340/2016-3, and 140466/2018-5), the São Paulo Research Foundation - FAPESP (grants 2013/08293-7, 2015/11937-9, 2017/12646-3, and 2017/16246-0), the Coordenação de Aperfeiçoamento de Pessoal de Nível Superior - Brasil (CAPES) - Finance Code 001, and the CAPES/COFECUB program (grant 831/15).

# References

1. Bafna, V., Pevzner, P.A.: Sorting by transpositions. SIAM J. Discrete Math. **11**(2), 224–240 (1998). https://doi.org/10.1137/S089548019528280X
2. Belda, E., Moya, A., Silva, F.J.: Genome rearrangement distances and gene order phylogeny in γ-proteobacteria. Mol. Biol. Evol. **22**(6), 1456–1467 (2005). https://doi.org/10.1093/molbev/msi134
3. Biller, P., Guéguen, L., Knibbe, C., Tannier, E.: Breaking good: accounting for fragility of genomic regions in rearrangement distance estimation. Genome Biol. Evol. **8**(5), 1427–1439 (2016). https://doi.org/10.1093/gbe/evw083
4. Biller, P., Knibbe, C., Beslon, G., Tannier, E.: Comparative genomics on artificial life. In: Beckmann, A., Bienvenu, L., Jonoska, N. (eds.) CiE 2016. LNCS, vol. 9709, pp. 35–44. Springer, Cham (2016). https://doi.org/10.1007/978-3-319-40189-8_4
5. Brito, K.L., Jean, G., Fertin, G., Oliveira, A.R., Dias, U., Dias, Z.: Sorting by genome rearrangements on both gene order and intergenic sizes. J. Comput. Biol. **27**(2), 156–174 (2020). https://doi.org/10.1089/cmb.2019.0293
6. Bulteau, L., Fertin, G., Rusu, I.: Sorting by transpositions is difficult. SIAM J. Comput. **26**(3), 1148–1180 (2012). https://doi.org/10.1137/110851390
7. Bulteau, L., Fertin, G., Tannier, E.: Genome rearrangements with indels in intergenes restrict the scenario space. BMC Bioinform. **17**(S14), 225–231 (2016). https://doi.org/10.1186/s12859-016-1264-6
8. Elias, I., Hartman, T.: A 1.375-approximation algorithm for sorting by transpositions. IEEE/ACM Trans. Comput. Biol. Bioinform. **3**(4), 369–379 (2006). https://doi.org/10.1109/TCBB.2006.44
9. Fertin, G., Jean, G., Tannier, E.: Algorithms for computing the double cut and join distance on both gene order and intergenic sizes. Algorithm Mol. Biol. **12**(16), 1–11 (2017). https://doi.org/10.1186/s13015-017-0107-y
10. Hannenhalli, S., Pevzner, P.A.: Transforming men into mice (polynomial algorithm for genomic distance problem). In: Proceedings of the 36th Annual IEEE Symposium Foundations of Computer Science (FOCS 1995), pp. 581–592. IEEE Computer Society Press, Washington, DC (1995). https://doi.org/10.1109/SFCS.1995.492588

11. Kececioglu, J.D., Sankoff, D.: Exact and approximation algorithms for sorting by reversals, with application to genome rearrangement. Algorithmica **13**, 180–210 (1995). https://doi.org/10.1007/BF01188586
12. Lin, Y., Rajan, V., Moret, B.M.E.: TIBA: a tool for phylogeny inference from rearrangement data with bootstrap analysis. Bioinformatics **28**(24), 3324–3325 (2012). https://doi.org/10.1093/bioinformatics/bts603
13. Oliveira, A.R., Jean, G., Fertin, G., Dias, U., Dias, Z.: Super short operations on both gene order and intergenic sizes. Algorithm Mol. Biol. **14**(21), 1–17 (2019). https://doi.org/10.1186/s13015-019-0156-5
14. Wang, L.S., Warnow, T., Moret, B.M.E., Jansen, R.K., Raubeson, L.A.: Distance-based genome rearrangement phylogeny. J. Mol. Evol. **63**(4), 473–483 (2006). https://doi.org/10.1007/s00239-005-0216-y

# Heuristics for Reversal Distance Between Genomes with Duplicated Genes

Gabriel Siqueira[1] , Klairton Lima Brito[1(✉)] , Ulisses Dias[2] ,
and Zanoni Dias[1]

[1] Institute of Computing, University of Campinas, Campinas, Brazil
gabriel.siqueira@students.ic.unicamp.br,
{klairton,zanoni}@ic.unicamp.br
[2] School of Technology, University of Campinas, Limeira, Brazil
ulisses@ft.unicamp.br

**Abstract.** In comparative genomics, one goal is to find similarities between genomes of different organisms. Comparisons using genome features like genes, gene order, and regulatory sequences are carried out with this purpose in mind.

Genome rearrangements are mutational events that affect large extensions of the genome. They are responsible for creating extant species with conserved genes in different positions across genomes.

Close species—from an evolutionary point of view—tend to have the same set of genes or share most of them. When we consider gene order to compare two genomes, it is possible to use a parsimony criterion to estimate how close the species are. We are interested in the shortest sequence of genome rearrangements capable of transforming one genome into the other, which is named *rearrangement distance*.

Reversal is one of the most studied genome rearrangements events. This event acts in a segment of the genome, inverting the position and possibly the orientation of genes in it.

When the genome has no gene repetition, a common approach is to map it as a permutation such that each element represents a conserved block.

When genomes have replicated genes, this mapping is usually performed using strings. The number of replicas depends on the organisms being compared, but in many scenarios, it tends to be small. In this work, we study the reversal distance between genomes with duplicated genes considering that the orientation of genes is unknown. We present three heuristics that use techniques like genetic algorithms and local search. We conduct experiments using a database of simulated genomes and compared our results with other algorithms from the literature.

**Keywords:** Genome rearrangement · Reversal · Heuristics · Duplicated genes

© Springer Nature Switzerland AG 2020
C. Martín-Vide et al. (Eds.): AlCoB 2020, LNBI 12099, pp. 29–40, 2020.
https://doi.org/10.1007/978-3-030-42266-0_3

# 1   Introduction

The question that naturally arises when comparing the genomes of two organisms is how to estimate the sequence of mutational events that have occurred during evolution to transform one genome into another, or at least estimate the evolutionary distance between these genomes.

A way of estimating the evolutionary distance is to use a parsimony criterion and to compute a minimum sequence of events that transforms one genome into another. When large-scale mutational events are considered, the so-called genome rearrangements, this distance is called *rearrangement distance.*

A genome can be represented in different ways [6]. When the genome is treated as an ordered sequence of genes, it is possible to find scenarios where certain genes have multiple copies. In this case, it is common to adopt a representation in the form of a string, such that each character is associated with a specific gene. If each gene occurs only once, we can associate an integer number for each one and the representation is given in the form of a permutation. In both cases (string or permutation), if the orientation of the genes is known, a positive or negative sign is assigned to each element and the representation is called signed (signed string and signed permutation). Otherwise, the sign is omitted and the representation is called unsigned (unsigned string and unsigned permutation).

Reversal is a rearrangement event that breaks a chromosome at two locations and reassembles the middle piece in the reversed order. Several rearrangement problems considering only reversals were investigated over time [1–3,8,11]. When representing genomes as permutations, the goal is to determine the minimum number of reversals needed to sort any permutation.

In this case, we have the Sorting Signed Permutations by Reversals and Sorting Unsigned Permutations by Reversals problems. The former is solvable in polynomial time [8], whereas the latter is NP-hard [3]. The best algorithm for the latter has an approximation factor of 1.375 [2].

In genomes with replicated genes, when we have the same number of replicas of each gene in both genomes, the goal is to determine the minimum number of reversals needed to transform one genome into another.

The Reversal Distance for Unsigned Strings problem is NP-hard even if we consider a binary alphabet [5]. A binary alphabet means that all genes are replicas of just two types, so the genome can be mapped using only two values (e.g. {0,1}).

Unlike Sorting Signed Permutations by Reversals problem which has an exact polynomial-time algorithm, the Reversal Distance for Signed Strings problem is NP-hard [15]. Chen *et al.* [4] proved that Reversal Distance for Signed Strings is NP-hard even if we consider the simplest case where at most two replicas are allowed for each gene (duplicated genes). The authors also showed that the Reversal Distance for Signed Strings and the Minimum Common String Partition (MCSP) problems are related. The Reversal Distance for Unsigned Strings and Reverse MCSP [11] problems are also related. Based on this information, an approximation algorithm for the Reversal Distance in Signed and Unsigned

Strings problems were presented with a factor of $\Theta(k)$, where $k$ represents the maximum number of copies of a character in the strings given as input to the algorithms [12].

In this paper, we investigate the Reversal Distance for Unsigned Strings problem considering duplicated genes. We propose three heuristics based on different techniques, and to verify the behavior of our heuristics we have created a database that simulates different scenarios. Our results were compared with others from the literature [4].

This manuscript is organized as follows. Section 2 provides definitions that are used throughout the paper. Section 3 describes the heuristics. Section 4 shows the experimental results, and Sect. 5 concludes the paper.

## 2   Basic Definitions

A genome $\mathcal{G}$ is represented by a string $S$, where each character in $S$ corresponds to a gene or block of genes in $\mathcal{G}$. An alphabet $\Sigma_S$ is the set of distinct characters of $S$. We denote by $S_i$ the $i$-th character in $S$, and by $|S|$ the number of characters.

*Example 1.* A string $S$ and some information we retrieve from it.

$$S = (5\ \ 2\ \ 1\ \ 3\ \ 4\ \ 5\ \ 4), \quad \Sigma_S = \{1, 2, 3, 4, 5\}, \quad |S| = 7, \ S_3 = 1\ , \ S_5 = 4.$$

**Definition 2.** *The occurrence of a character $\alpha$ in a given string $S$, denoted by $occ(\alpha, S)$, represents the number of copies of $\alpha$ in $S$. The greatest occurrence of a character in $S$ is denoted by $occ(S) = \max_{\alpha \in \Sigma_S}(occ(\alpha, S))$.*

**Definition 3.** *We denote by $dup(S)$ the set of duplicated characters of $S$, i.e. the characters that appear exactly twice in $S$. Therefore, $dup(S) = \{\alpha : occ(\alpha, S) = 2, \forall \alpha \in S\}$.*

In Example 1, we have $occ(S) = 2$, and $dup(S) = \{4, 5\}$.

**Definition 4.** *A pair of strings $S$ and $P$ are balanced if they have the same alphabet ($\Sigma_S = \Sigma_P = \Sigma$) and the occurrence of each character is the same for both strings. Therefore, $occ(\alpha, S) = occ(\alpha, P)$, $\forall\ \alpha \in \Sigma$.*

*Example 5.* Consider three strings $S$, $P$, and $Q$. Observe that $S$ and $P$ are balanced while $S$ and $Q$ are not, since the occurrences of character 1 in the strings $S$ and $Q$ are different ($occ(1, S) \neq occ(1, Q)$).

$$S = (5\ \ 2\ \ 1\ \ 3\ \ 4\ \ 5\ \ 4)$$
$$P = (4\ \ 4\ \ 1\ \ 2\ \ 5\ \ 5\ \ 3)$$
$$Q = (5\ \ 1\ \ 1\ \ 3\ \ 4\ \ 5\ \ 4\ \ 2)$$
$$\Sigma_S = \Sigma_P = \Sigma_Q$$

We represent genome rearrangement events as operations applied to strings. This way, a rearrangement event $\rho$ applied to a string $S$ is denoted as $S \circ \rho$.

**Definition 6.** *A reversal $\rho(i,j)$, with $1 \leq i < j \leq |S|$, is an operation that inverts the order of elements in a segment of the string $S$.*

$$S \qquad = (S_1 \ldots S_{i-1} \; \underline{S_i \ldots S_j} \; S_{j+1} \ldots S_{|S|})$$
$$S \circ \rho(i,j) = (S_1 \ldots S_{i-1} \; \underline{S_j \ldots S_i} \; S_{j+1} \ldots S_{|S|})$$

*Example 7.* A reversal $\rho(2,4)$ applied on a string $S$.

$$S \qquad = (1 \; \underline{2 \; 3 \; 3} \; 2 \; 1 \; 4)$$
$$S \circ \rho(2,4) = (1 \; \underline{3 \; 3 \; 2} \; 2 \; 1 \; 4)$$

**Definition 8.** *Given two strings $S$ and $P$, the reversal distance between $S$ and $P$, denoted by $d(S,P)$, is the size of a shortest sequence of reversals capable of transforming $S$ into $P$.*

From now on, we refer to the reversal distance only by distance. This work deals with balanced strings $S$ such that $occ(S) \leq 2$. Every genome with multiple copies of a gene can be mapped into a string. We step forward and map the string into a permutation. To do that, we keep the characters without duplicates untouched and map each replica of duplicated characters into new values.

Assuming two replicas of a character $\alpha$, there are two possible mappings of $\alpha$ into new values $\alpha'$ and $\alpha''$. The mapping of all duplicated characters in $S$ is represented by a vector $\mathbf{m}$ with size $|dup(S)|$. In $\mathbf{m}$ we place the value 0 or 1 to indicate for each duplicated character which of the two possible maps will be used. If the value associated with the character $\alpha$ in $\mathbf{m}$ is 0, the first and the second occurrences will be replaced by $\alpha'$ and $\alpha''$, respectively. Otherwise, the first and second occurrences will be mapped as $\alpha''$ and $\alpha'$, respectively.

We denote by $\mathbf{m}_\alpha$ the chosen map of the duplicate character $\alpha$ in $\mathbf{m}$ and by $S^{\mathbf{m}}$ the permutation generated by mapping $S$ according to $\mathbf{m}$.

*Example 9.* A map $\mathbf{m}$ being applied to a string $S$ and the permutation $S^{\mathbf{m}}$ obtained.

$$S \; = (5 \; 2 \; 1 \; 3 \; 4 \; 5 \; 4), \quad dup(S) = \{4,5\}$$
$$S^{\mathbf{m}} = (5'' \; 2 \; 1 \; 3 \; 4' \; 5' \; 4'')$$

$$\overset{4 \; 5}{\mathbf{m} = \boxed{0\,1}}, \quad \mathbf{m}_4 = 0, \quad \mathbf{m}_5 = 1$$

**Definition 10.** *Given a string $S$ and two maps $\mathbf{m}$ and $\mathbf{v}$. We say that $\mathbf{m}$ and $\mathbf{v}$ are neighbors if they differ by exactly one duplicated character map. In other words, $\exists \alpha \in dup(S) : \mathbf{m}_\alpha \neq \mathbf{v}_\alpha$ and $\mathbf{m}_\beta = \mathbf{v}_\beta, \forall_{\beta \neq \alpha}$.*

*Example 11.* We obtain $S^{\mathbf{m}}$, $S^{\mathbf{v}}$, and $S^{\mathbf{z}}$ from $S$ using maps $\mathbf{m}$, $\mathbf{v}$, and $\mathbf{z}$, respectively. Note that $(\mathbf{m}, \mathbf{v})$ and $(\mathbf{v}, \mathbf{z})$ are neighbors, but $(\mathbf{m}, \mathbf{z})$ are not.

$$
\begin{aligned}
S \; &= (1 \; 2 \; 1 \; 3 \; 3 \; 2), \quad dup(S) = \{1, 2, 3\} \\
S^{\mathbf{m}} &= (1' \; 2' \; 1'' \; 3'' \; 3' \; 2'') \\
S^{\mathbf{v}} &= (1' \; 2' \; 1'' \; 3' \; 3'' \; 2'') \\
S^{\mathbf{z}} &= (1'' \; 2' \; 1' \; 3' \; 3'' \; 2'')
\end{aligned}
$$

$$
\mathbf{m} = \begin{array}{|c|c|c|} \hline 0 & 0 & 1 \\ \hline \end{array} \quad
\mathbf{v} = \begin{array}{|c|c|c|} \hline 0 & 0 & 0 \\ \hline \end{array} \quad
\mathbf{z} = \begin{array}{|c|c|c|} \hline 1 & 0 & 0 \\ \hline \end{array}
$$

(column labels $1\ 2\ 3$ above each)

## 3 Heuristic Approaches

Although the task that models genomes as permutations does not allow duplicated genes, we can perform a mapping of the strings into permutations by assigning new values to replicas. We base our heuristics on the fact that if we obtain two permutations $S^{\mathbf{m}}$ and $P^{\mathbf{p}}$ from two strings $S$ and $P$ using the maps $\mathbf{m}$ and $\mathbf{p}$, respectively, then the sequence of reversals that turns $S^{\mathbf{m}}$ into $P^{\mathbf{p}}$ also transforms $S$ into $P$. Therefore, $d(S, P) \leq d(S^m, P^p)$.

Note that, there exist maps such that $d(S, P) = d(S^m, P^p)$. Such maps could be derived from a shortest sequence of reversals transforming S in P.

*Example 12.* A sequence of reversals transforming $S^m$ into $P^p$ and $S$ into $P$.

$$
\begin{aligned}
S^m &= (5'' \; \underbrace{2 \; 1}_{\rho(2,3)} \; 3 \; 4' \; 5' \; 4'') \\
&= (5'' \; 1 \; 2 \; \underbrace{3 \; 4' \; 5' \; 4''}_{\rho(4,7)}) \\
&= (5'' \; 1 \; \underbrace{2 \; 4'' \; 5'}_{\rho(3,5)} \; 4' \; 3) \\
P^p &= (5'' \; 1 \; 5' \; 4'' \; 2 \; 4' \; 3)
\end{aligned}
\qquad
\begin{aligned}
S &= (5 \; \underbrace{2 \; 1}_{\rho(2,3)} \; 3 \; 4 \; 5 \; 4) \\
&= (5 \; 1 \; 2 \; \underbrace{3 \; 4 \; 5 \; 4}_{\rho(4,7)}) \\
&= (5 \; 1 \; \underbrace{2 \; 4 \; 5}_{\rho(3,5)} \; 4 \; 3) \\
P &= (5 \; 1 \; 5 \; 4 \; 2 \; 4 \; 3)
\end{aligned}
$$

$$
\mathbf{m} = \begin{array}{|c|c|} \hline 0 & 1 \\ \hline \end{array} \quad
\mathbf{p} = \begin{array}{|c|c|} \hline 1 & 1 \\ \hline \end{array}
$$

(column labels $4\ 5$ above each)

To find the distance between two strings we must find the map of the strings in the permutations with the smallest distance. We use as a distance estimator for the Sorting Unsigned Permutations by Reversals problem an approximation algorithm with factor 2 developed by Kececioglu and Sankoff [10], which we will call KS95. We chose this algorithm because its results in practice are good and its execution time is fast, which serves our purpose of creating simple heuristics that provide solutions in a fast way. However, another much more complicated algorithm with a better approximation ratio is known [2].

Our heuristics share a common goal: find a map of strings into permutations that results in a good solution. Sections 3.1, 3.2, and 3.3 present heuristics using Random Maps, Local Search, and Genetic Algorithms, respectively.

## 3.1  Random Maps (RM)

This heuristic randomly generates several maps of the source string into permutations, and a single random map of the target string. After that, the heuristic estimates the distance between each source permutation and the target permutation using KS95. In the end, the heuristic selects the solution with smallest distance. In case of a tie, the heuristic selects one of the best solutions randomly.

The inputs are the strings $S$ and $P$, and a parameter $r \in \mathbb{N}$ that define the total number of random maps for the source string. The random maps are generated as follows: for each position of the map, the values 0 and 1 are assigned with same probability. Initially, the target string $P$ is mapped into a permutation $P^\mathbf{p}$ using a random map $\mathbf{p}$. Next, $r$ random maps of the source string $S$ are generated and stored in a set $\mathbf{M}$. For each map $\mathbf{m} \in \mathbf{M}$, the distance between the permutations $S^\mathbf{m}$ and $P^\mathbf{p}$ is computed using KS95. The result for $S$ and $P$ is the shortest distance between the permutations.

Figure 1 shows a simulation of the heuristic for $S = (3\ 2\ 1\ 2\ 4\ 3\ 4)$ and $P = (1\ 3\ 4\ 2\ 2\ 4\ 3)$. The heuristic finds that $d(S, P) \leq d(S^m, P^p) \leq 4$.

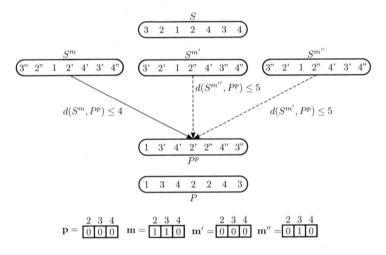

**Fig. 1.** Example of random maps heuristic with $r = 3$.

The random mapping process described in this heuristic will be used as subroutines by other heuristics.

## 3.2  Local Search (LS)

This heuristic enhances the Random Maps by generating maps through local search technique and not exclusively in a random way. The idea is to create new maps by selecting some of the current ones to have their neighborhoods explored.

Inputs are strings $S$ and $P$, and the parameters $r, c, l \in \mathbb{N}$. Parameters $r$, $c$, and $l$ determine the total number of maps that will be created, the number of maps randomly generated, and the maximum number of neighbors explored in each local search, respectively.

Initially, the heuristic behaves like the Random Maps heuristic: (i) the target string $P$ is mapped into a permutation $P^P$ using a random map $\mathbf{p}$, and (ii) a set $\mathbf{M}$ is generated with $c$ random maps of the source string $S$ into permutations. Note that $\mathbf{M}$ is composed of only $c$ randomly generated maps, but this set must contain $r$ distinct maps. To generate $r - c$ maps the heuristic performs a local search on the best solutions found so far as follows:

1. In each iteration, the heuristic ranks the maps using the algorithm KS95.
2. The heuristic selects the best map and explores up to $l$ neighbor maps, adding them to $\mathbf{M}$. This process ends when the set $\mathbf{M}$ has $r$ distinct maps.
3. The heuristic keeps a list of the maps that have already been explored. Thus, it ensures that a map is explored only once. This behavior is important to explore the neighborhood of other maps that are also good.

Note that a map created through local search in one iteration can have the neighborhood explored in future iterations.

Similarly to Random Maps heuristic, the result for the distance between the strings $S$ and $P$ is the shortest distance calculated between the permutations resulting from maps in $\mathbf{M}$ and the permutation $P^P$.

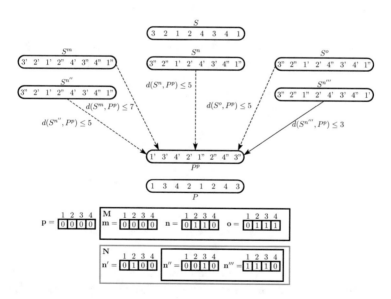

**Fig. 2.** Example of local search heuristic with $r = 5$, $c = 3$, and $l = 2$.

Figure 2 illustrates this heuristic in a pair of strings $S = (3\ 2\ 1\ 2\ 4\ 3\ 4\ 1)$ and $P = (1\ 3\ 4\ 2\ 1\ 2\ 4\ 3)$. The initial set $\mathbf{M} = \{\mathbf{m}, \mathbf{n}, \mathbf{o}\}$ is compose of 3 randomly generated maps. From $\mathbf{M}$, the map $\mathbf{n}$ is selected to have the neighborhood explored. Set $\mathbf{N} = \{\mathbf{n}', \mathbf{n}'', \mathbf{n}'''\}$ represents all neighboring maps of $\mathbf{n}$ that are not yet in $\mathbf{M}$. Adopting the parameter $l = 2$, the heuristic chooses the maps $\mathbf{n}''$ and $\mathbf{n}'''$ to add into $\mathbf{M}$, filling the set with $r = 5$ distinct maps. In that case, the heuristic finds that $d(S, P) \le d(S^{n'''}, P^p) \le 3$.

### 3.3   Genetic Algorithm (GA)

This heuristic is modeled using Genetic Algorithm metaheuristic [13] to generate the maps. This meta-heuristic is widely used on combinatorial optimization problems [9,14] and has already been used in problems of genome rearrangement [7].

A genetic algorithm is a search heuristic inspired by the theory of evolution. It uses features like mutations, inheritance of parents characteristics, and selection of fittest individuals for reproduction, to name a few.

The inputs are the strings $S$ and $P$, and the parameters $c, k, r, t_m, t_c \in \mathbb{N}$. The parameter $c$ is the initial population size, $k$ is the number of individuals selected in a given population, and $r$ is the total number of maps created for the source string. The parameters $t_m$ and $t_c$ are used in the mutations and crossovers, respectively. The population size, in each generation, is $\frac{5k}{2}$. Once $r$ is reached, the algorithm stops and the best result so far is returned.

We describe our genetic algorithm considering five features: initial population, fitness function, selection, crossover, and mutation.

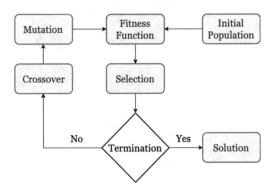

**Fig. 3.** Flowchart of genetic algorithm phases.

Figure 3 shows a flowchart of the interactions between each phase to obtain a solution using the genetic algorithm technique. We developed a heuristic where each individual in the population is represented by the mapping of the source string $S$ into a permutation. We also used a single random map $\mathbf{p}$ of the target string $P$ into a permutation $P^p$.

- **Initial Population:** is generated $c$ random maps of the source string S into distinct permutations. The process of generating random maps is the same as that used in the Random Maps heuristic.

- **Fitness Function:** each individual receives a score based on a fitness function. The fitness function assigns a score to an individual **m** of the population as follows: $\mathbf{m}_{score} = \frac{1}{d(S^{\mathbf{m}}, P^{\mathbf{P}})+1}$, such that $d(S^{\mathbf{m}}, P^{\mathbf{P}})$ is computed using the KS95 algorithm. Maps that result in solutions with smaller number of reversals receive higher scores and tend to transmit their characteristics to future generations.
- **Selection:** at this stage, the top $k$ highest scored individuals are selected to the next generation, and the others are discarded. In cases of ties, the heuristic arbitrarily selects the required number of individuals among the tied ones.
- **Crossover:** at this phase, the population has the $k$ highest scored individuals. To perform the re-population, individuals are arbitrarily paired for crossovers. Given two maps **m** and **m'** with $x = dup(S) = dup(P)$ bits, the crossover generates a new individual with $t_c$ bits randomly selected from **m** and the remaining $x - t_c$ from **m'**.

*Example 13.* Maps **m** and **m'** of a string $S = (3\ 3\ 1\ 4\ 2\ 2\ 4)$ adopting $t_c = 2$ and resulting in a new individual **n**.

$$
\begin{aligned}
S &= (3\ 3\ 1\ 4\ 2\ 2\ 4), \quad t_c = 2 \\
S^m &= (3''\ 3'\ 1\ \underline{4''\ 2'\ 2''\ 4'}) \\
S^{m'} &= (\underline{3'\ 3''}\ 1\ 4'\ 2''\ 2'\ 4'') \\
S^n &= (\underline{3'\ 3''}\ 1\ \underline{4''\ 2'\ 2''\ 4'})
\end{aligned}
$$

$$
\mathbf{m} = \begin{array}{ccc} 2 & 3 & 4 \\ \hline 0 & 1 & 1 \end{array} \quad \mathbf{m'} = \begin{array}{ccc} 2 & 3 & 4 \\ \hline 1 & 0 & 0 \end{array} \quad \mathbf{n} = \begin{array}{ccc} 2 & 3 & 4 \\ \hline 0 & 0 & 1 \end{array}
$$

- **Mutation:** after crossover, each of the $k$ individuals selected from the past generation gives rise to a new individual by inverting the values of $t_m$ bits chosen randomly.

*Example 14.* A map **n** generated by a mutation in a map **m** of a string $S = (3\ 3\ 1\ 4\ 2\ 2\ 4)$ adopting $t_m = 2$.

$$
\begin{aligned}
S &= (3\ 3\ 1\ 4\ 2\ 2\ 4), \quad t_m = 2 \\
S^m &= (3''\ 3'\ 1\ 4''\ 2'\ 2''\ 4') \\
S^n &= (3'\ 3''\ 1\ 4'\ 2'\ 2''\ 4'')
\end{aligned}
$$

$$
\mathbf{m} = \begin{array}{ccc} 2 & 3 & 4 \\ \hline 0 & 1 & 1 \end{array} \quad \mathbf{n} = \begin{array}{ccc} 2 & 3 & 4 \\ \hline 0 & 0 & 0 \end{array}
$$

The heuristic repeats this process until $r$ maps are generated. Afterwards, the individual **m** with the highest score is selected to obtain a solution computing the distance between $S^{\mathbf{m}}$ and $P^{\mathbf{P}}$ using the KS95 algorithm.

## 4    Experimental Results

We present our test methodology and the results obtained by our heuristics. In the end, we compare our results with others from the literature.

## 4.1  Database

Our database comprises 10 sets of 1000 pairs of strings (source and target). Each set has strings of different sizes ranging from 100 to 1000 in intervals of 100. Strings have 25% of duplicated characters, so $|dup(S)| = \frac{|S|}{4}$. Each pair of source and target strings was created as follows: we randomly distributed the values $\{1, 2, \ldots, \frac{3|S|}{4}, 1, 2, \ldots, \frac{|S|}{4}\}$ to create the source string $S$, and we applied a total of $\frac{|S|}{4}$ random reversals (the parameters $i$ and $j$ being chosen randomly) to generate the target string $P$.

## 4.2  Model Tuning

Random Maps, Local Search, and Genetic Algorithm heuristics share a common $r$ parameter, which represents the total number of maps that are generated. This parameter should have the same number set for us to be able to compare the heuristics. In other words, heuristics must explore the same number of maps, so we may estimate the gain of each model in choosing good maps to investigate. We assigned $r = 10|S|$ for each source string $S$ during the experiments.

Other parameters of Local Search and Genetic Algorithm heuristics were selected using a grid search. For Local Search heuristic parameters $c$ and $l$, we swept through the set $\{10, 20, \ldots, 100\}$. For the Genetic Algorithm heuristic, $k$ was investigated in $\{10, 20, \ldots, 100\}$, $c$ in $\{10, 20, \ldots, 100\}$, $t_m$ in $\{1, 2, \ldots, 10\}$, and $t_c$ in $\{\lfloor 0.1d \rfloor, \lfloor 0.2d \rfloor, \ldots, d\}$, where $d = |dup(S)| = |dup(P)|$. The grid search was performed in the sets of strings of sizes 400, 500, and 600. We obtained the following parameters:

– Local Search: $c = 90$ and $l = 30$;
– Genetic Algorithm: $c = 90$, $k = 50$, $t_m = 2$, and $t_c = \lfloor 0.4d \rfloor$.

## 4.3  Results

For comparison purposes, we implemented the Kolman and Waleń algorithm [12] (called **HS**) and an adaptation of the **SOAR** algorithm [4]. This adaptation was performed to consider the Reverse MCSP [11] instead of MCSP problem [4] in order to address the unsigned version of the Reversal Distance for Strings problem.

The abbreviations **RM**, **LS**, and **GA** refer to Random Maps, Local Search, and Genetic Algorithm heuristics, respectively. Table 1 shows the average distance provided by the heuristics, **HS**, and **SOAR** using our database as input.

The **OP** column shows the number of random reversals used to create each instance. The line (**DEE**$_{\mathbf{avg}}$) represents the average distance estimation error. For an instance $(S, P)$, the distance estimation error is calculated as follows: $\frac{|D_H - OP|}{OP}$, such that $D_H$ is the distance estimation for the instance $(S, P)$ computed by our heuristics and **SOAR** algorithm. The distance estimation error shows, as a percentage of the number of reversals applied to create the instance,

**Table 1.** Results provided by our heuristics and by **SOAR**.

| String size | RM | LS | GA | SOAR | HS | OP |
|---|---|---|---|---|---|---|
| 100 | 32.62 | 24.22 | 24.33 | 29.65 | 50.27 | 25 |
| 200 | 75.31 | 49.61 | 49.75 | 61.42 | 104.71 | 50 |
| 300 | 120.94 | 75.39 | 75.60 | 94.16 | 160.41 | 75 |
| 400 | 167.30 | 100.99 | 101.30 | 126.34 | 214.97 | 100 |
| 500 | 214.36 | 126.75 | 126.89 | 158.97 | 270.93 | 125 |
| 600 | 262.34 | 152.74 | 152.91 | 191.21 | 326.75 | 150 |
| 700 | 310.99 | 179.64 | 179.17 | 224.11 | 382.77 | 175 |
| 800 | 359.60 | 207.73 | 205.33 | 256.41 | 438.87 | 200 |
| 900 | 408.31 | 238.63 | 231.81 | 289.04 | 494.33 | 225 |
| 1000 | 457.22 | 274.03 | 258.54 | 321.88 | 549.91 | 250 |
| $DEE_{avg}$ | 67.81% | 3.87% | 2.88% | 26.20% | 115.17% | – |

how far the heuristics and **SOAR** algorithm have distanced from **OP**, either for more or less.

From the results, we can see that although the **HS** algorithm guarantees an approximation to the solution by a multiplicative factor $\Theta(k)$, in practice, the results of the other algorithms were better. This result was probably caused by the fact that the constant associated with function $\Theta(k)$ has a high value.

Genetic Algorithm (**GA**) and Local Search (**LS**) are better than **SOAR** and **RM** in all sets, which clearly indicates that the more sophisticated rules added in **GA** and **LS** succeed. On average, these strategies provide values very close to the number of reversals applied to generate the instances. This can be evidenced by observing the average distance estimation error (**DEE$_{avg}$**) which, considering all instances, shows **GA** and **LS** heuristics with a percentage of 2.88% and 3.87%, respectively. We also can note that **GA** generates the best distance estimator (closest to **OP**) for all sets tested.

The heuristics **LS** and **GA** presented a far better performance than the previous know methods for estimate the evolutionary distance between genomes with duplicated genes, considering the reversal event.

We also performed tests with different values for the parameter $r$ ranging from $5|S|$ up to $50|S|$, but the best trade-off between solution quality and runtime was observed adopting $r = 10|S|$.

## 5   Conclusion

We presented three heuristics for the Reversal Distance for Unsigned Strings with Duplicated Genes problem. We performed experiments with a database created to simulate genomes with different characteristics. We compared the results obtained by our heuristics with results from the literature. The comparison shows that our heuristic based on local search and genetic algorithms techniques tend to produce very good solutions.

As future works, we plan to extend the heuristics by considering other genome rearrangement events (e.g., transposition, insertion, and deletion), other meta-heuristics, and by investigating the problem having more than two copies for each character.

**Acknowledgments.** This work was supported by the National Council for Scientific and Technological Development - CNPq (grants 400487/2016-0, 425340/2016-3, 304380/2018-0, and 140466/2018-5), the São Paulo Research Foundation - FAPESP (grants 2015/11937-9, 2017/12646-3, and 2017/16246-0), the Brazilian Federal Agency for the Support and Evaluation of Graduate Education - CAPES, and the CAPES/COFECUB program (grant 831/15).

# References

1. Bergeron, A.: A very elementary presentation of the Hannenhalli-Pevzner theory. Discrete Appl. Math. **146**(2), 134–145 (2005)
2. Berman, P., Hannenhalli, S., Karpinski, M.: 1.375-Approximation algorithm for sorting by reversals. In: Möhring, R., Raman, R. (eds.) ESA 2002. LNCS, vol. 2461, pp. 200–210. Springer, Heidelberg (2002). https://doi.org/10.1007/3-540-45749-6_21
3. Caprara, A.: Sorting permutations by reversals and Eulerian cycle decompositions. SIAM J. Discrete Math. **12**(1), 91–110 (1999)
4. Chen, X., Zheng, J., Fu, Z., Nan, P., Zhong, Y., Lonardi, S., Jiang, T.: Assignment of orthologous genes via genome rearrangement. IEEE/ACM Trans. Comput. Biol. Bioinform. **2**(4), 302–315 (2005)
5. Christie, D.A., Irving, R.W.: Sorting strings by reversals and by transpositions. SIAM J. Discrete Math. **14**(2), 193–206 (2001)
6. Fertin, G., Labarre, A., Rusu, I., Tannier, É., Vialette, S.: Combinatorics of genome rearrangements. Computational Molecular Biology. The MIT Press, London (2009)
7. Gao, N., Yang, N., Tang, J.: Ancestral genome inference using a genetic algorithm approach. PLOS ONE **8**(5), 1–6 (2013)
8. Hannenhalli, S., Pevzner, P.A.: Transforming cabbage into turnip: polynomial algorithm for sorting signed permutations by reversals. J. ACM **46**(1), 1–27 (1999)
9. Jog, P., Suh, J., Van Gucht, D.: Parallel genetic algorithms applied to the traveling salesman problem. SIAM J. Optimization **1**, 515–529 (1991)
10. Kececioglu, J.D., Sankoff, D.: Exact and approximation algorithms for sorting by reversals, with application to genome rearrangement. Algorithmica **13**, 180–210 (1995)
11. Kolman, P., Waleń, T.: Approximating reversal distance for strings with bounded number of duplicates. Discrete Appl. Math. **155**(3), 327–336 (2007)
12. Kolman, P., Waleń, T.: Reversal distance for strings with duplicates: linear time approximation using hitting set. In: Erlebach, T., Kaklamanis, C. (eds.) WAOA 2006. LNCS, vol. 4368, pp. 279–289. Springer, Heidelberg (2007). https://doi.org/10.1007/11970125_22
13. Mitchell, M.: An Introduction to Genetic Algorithms. MIT Press, Cambridge (1996)
14. Pezzella, F., Morganti, G., Ciaschetti, G.: A genetic algorithm for the flexible job-shop scheduling problem. Comput. Oper. Res. **35**(10), 3202–3212 (2008)
15. Radcliffe, A.J., Scott, A.D., Wilmer, E.L.: Reversals and transpositions over finite alphabets. SIAM J. Discrete Math. **19**(1), 224–244 (2005)

# Extending Maximal Perfect Haplotype Blocks to the Realm of Pangenomics

Lucia Williams$^{(\boxtimes)}$ and Brendan Mumey

Montana State University, Bozeman, MT 59718, USA
lucia.williams@montana.edu

**Abstract.** Recent work provides the first method to measure the relative fitness of genomic variants within a population that scales to large numbers of genomes. A key component of the computation involves finding conserved haplotype blocks, which can be done in linear time. Here, we extend the notion of conserved haplotype blocks to pangenomes, which can store more complex variation than a single reference genome. We define a *maximal perfect pangenome haplotype block* and give a linear-time, suffix tree based approach to find all such blocks from a set of pangenome haplotypes. We demonstrate the method by applying it to a pangenome built from yeast strains.

**Keywords:** Population genomics · Haplotype block · Pangenomics

## 1 Introduction

Given the availability of sequenced genome data for many individuals of the same species, it is now possible to study population genetics and evolution at a level of detail not before possible. An established method for quantifying the relative fitness of two genetic variants uses the *selection coefficient* [6, Chapter 5.3]. Recent work by Cunha et al. [4] describes a method to scale the computation of selection coefficients across an entire genome, even when the number of individuals being analyzed is large. They adopt the maximum-likelihood based method from Chen et al. [3] for computing the selection coefficient for a maximally conserved portion of the genome. These conserved portions of the genome can be identified using *haplotypes*: sequences of single nucleotide polymorphism (SNP) sites defined with respect to a reference sequence for the population. However, Cunha et al. note that, prior to their work, no efficient method existed to compute all maximally conserved blocks from a set of haplotypes. They give an algorithm for locating the blocks that is quadratic in the length of the haplotypes. More recently, Alanko et al. [1] give a method for finding haplotype blocks in linear time. However, both haplotype block location algorithms assume that all genomes under consideration have been aligned to the same reference genome.

© Springer Nature Switzerland AG 2020
C. Martín-Vide et al. (Eds.): AlCoB 2020, LNBI 12099, pp. 41–48, 2020.
https://doi.org/10.1007/978-3-030-42266-0_4

A *pangenome* allows us to consider more complex variation in multiple individuals or organisms from a related group or species [10]. Pangenomic sequence data are often studied using graphs, where each sequence in a data set is represented by a path in the graph. In this work, we reformulate the problem of finding maximal haplotype blocks in the context of pangenomics. We give a method for finding pangenome SNPs in a De Bruijn graph in Sect. 3, define the pangenome maximal perfect haplotype block problem in Section 4, and describe a suffix tree approach to find all blocks in linear time relative to the input in Sect. 5. Finally, we find maximal perfect pangenome haplotype blocks in a ten-strain yeast pangenome and report results in Sect. 6.

## 2   Background

Given a set of binary sequences representing the presence (or absence) of SNPs in a chromosome, the authors of [4] define a *maximal perfect haplotype block* as follows:

**Definition 1.** *Given k sequences $S = (s_1, s_2, \ldots, s_k)$ of length n, a* maximal perfect haplotype block *is a triple (K, i, j) with $K \subseteq \{1, 2, \ldots, k\}$, $|K| \geq 2$, and $1 \leq i \leq j \leq n$ such that*

1. $s[i, j] = t[i, j]$ for all $s, t \in S|_K$ (equality),
2. $i = 1$ or $s[i-1] \neq t[i-1]$ for some $s, t \in S|_K$ (left-maximality),
3. $j = n$ or $s[j+1] \neq t[j+1]$ for some $s, t \in S|_K$ (right-maximality),
4. $\nexists K' \subseteq \{1, 2, \ldots, k\}$ with $K' \subsetneq K$ such that $s[i, j] = t[i, j]$ for all $s, t \in S|_{K'}$ (row-maximality).

Then, the *maximal perfect haplotype block* (MPHB) problem is to find all maximal perfect haplotypes in a given set of sequences. For example, Fig. 1 shows a set of three sequences containing five MPHBs.

In the case of pangenomic data it may not be possible to align each chromosome to a reference so we consider a generalized setting of the problem in which the SNPs occur in an arbitrary directed graph, rather than a linear sequence.

> Sequence 1: 1 0 1 1 1 1
> Sequence 2: 0 1 0 0 1 0
> Sequence 3: 1 0 0 0 1 0

**Fig. 1.** The five maximal perfect haplotype blocks in this set of sequences are $(\{1, 3\}, 1, 2), (\{1, 3\}, 2, 2), (\{1, 2, 3\}, 3, 4), (\{2, 3\}, 3, 6)$, and $(\{1, 2, 3\}, 5, 5)$.

## 3   Building the SNP Graph

We assume that a compressed De Bruijn graph (cDBG) has been built for the pangenomic data set we wish to study [2]. In this case the data set consists of a set of pangenomic sequences and the cDBG graph $G$ consists of a set of nodes representing specific $k$-mers (or $\geq k$-mers if the graph has been compressed). The parameter $k$ must be specified. An edge $(u, v)$ is present in $G$ provided the last $k - 1$ nucleotides of $u$ match the first $k - 1$ nucleotides of $v$. Each pangenomic sequence is associated with a path in $G$, where each path node appends all non-overlapping characters from the previous node in the path. Let $P$ denote the collection of sequence paths in $G$.

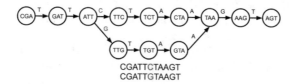

CGATTCTAAGT
CGATTGTAAGT

**Fig. 2.** A bubble in a De Bruijn graph that represents a SNP; we arbitrarily consider one side of the bubble to be the '0' path and the other to be the '1' path. In the compressed De Bruijn graph, the '0' and '1' paths are each a single node.

We identify *pangenomic SNPs* by looking for "bubbles" in $G$. Bubbles, as shown in Fig. 2, occur when paths diverge into exactly two subpaths and then rejoin, and no additional edges enter or leave the interior of the bubble. We view one side of the bubble as a '0' and the other as a '1'. Some bubbles will be longer than one nucleotide, but we still refer to them as SNPs for simplicity of notation. All SNPs can be found in $O(|G|)$ time, since bubble nodes in a cDBG can be recognized in $O(1)$ time. We form the SNP graph by retaining only those vertices of the cDBG graph that correspond to the '0' and '1' branches for each identified SNP. The paths $P$ in $G$ induce new SNP paths by deleting the non-SNP nodes in each path. The resulting SNP path sequences are used as input to maximal perfect perfect pangenome haplotype block problem, defined in the next section.

## 4   Problem Definition

Given a SNP graph and a sequence, a pangenome haplotype is the list of nodes that the sequence follows through the SNP graph. Due to large structural variations such as strain-specific genes, segmental deletions, insertions, and rearrangements, certain regions of the pangenome may be missed by some sequences but followed by others. Thus, not all pangenome haplotypes have the exact same set of SNPs, and the position of a node within the path does not indicate which SNP the node corresponds to as it does in the single-reference case. Instead, the node labels indicate both the SNP identifier and the call (either a '0' or a '1'). Figure 3 lists four example pangenome haplotypes.

We define a *maximal perfect pangenome haplotype block*.

**Definition 2.** *Given a set of k paths $P = (p_1, p_2, \ldots, p_k)$ through graph $G = (V, E)$, where each path is a sequence of nodes in $V$, a maximal perfect pangenome haplotype block is a set $K \subseteq \{1, 2, \ldots, k\}$ and a path of m nodes s such that:*

1. *s is a subpath of $p_i$ for all $i \in K$ (equality),*
2. *There is no in-neighbor u of $s[1]$ such that $u, s$ is a subpath of $p_i$ for all $i \in K$ (left maximality),*
3. *There is no out-neighbor v of $s[m]$ such that $s, v$ is a subpath of $p_i$ for all $i \in K$ (right maximality),*
4. *There is no $K' \subseteq \{1, 2, \ldots, k\}$ such that $K' \subsetneq K$ and s is a subpath of $p_i$ for all $i \in K'$ (path set maximality).*

Just as in the standard MPHB problem, the *maximal perfect pangenome haplotype block (MPPHB) problem* is to find all maximal perfect pangenome haplotype blocks among the $k$ paths.

We note that if $n$ is the length of the longest path in $P$, then there are no more than $(n + 1)k$ MPPHBs in any set of paths $P$. A proof is given in Sect. 5.

## 5   Linear Time Method Based on Suffix Trees

As in [1], we can use a suffix tree to solve the MPPHB problem in linear time.

Alanko et al. [1] note that all MPHBs in a set of sequences $S = \{s_1, s_2, \ldots, s_k\}$ correspond to *maximal repeats* (repeated substrings that cannot be extended; see [7, Section 7.12]) in the string $\mathbb{S} = s_1 \$_1 s_2 \$_2 \ldots s_k \$_l$. However, not all maximal repeats in $\mathbb{S}$ are MPHBs, since any $s_i$ may contain repeated substrings and a pair $s_i$ and $s_j$ may contain the same substring beginning at different positions. Neither of these is a MPHB.

They propose adding $n + 1$ unique "index characters" to each sequence, alternating with the existing characters. This way, substrings can only match to other substrings if they occur in exactly the same position in two different sequences. This process creates the string $\mathbb{S}^+$ so that there is a maximal repeat in $\mathbb{S}^+$ if and only if there is a MPHB in $S$. It is possible to find all maximal repeats in a string using a suffix tree in linear time and space [7, Section 7.12].

In the pangenome case, the suffix tree approach can still be applied. Because haplotype blocks need not begin at the same position in the path, the index characters are not needed. If the SNP graph contains cycles, then there may be

---

SNP sequence 1: [1:0, 2:0, 3:1, 6:0, 5:0, 10:1]
SNP sequence 2: [2:0, 3:1, 5:0, 6:1, 7:0, 8:0, 9:0]
SNP sequence 3: [1:1, 2:1, 6:1, 7:0, 8:0, 9:1, 10:1]

---

**Fig. 3.** Three pangenome sequences represented as paths through a SNP graph containing ten SNPs. The subpath [2:0 3:1] and sequences {1, 2} represent one maximal perfect pangenome haplotype block. Subpath [6:1, 7:0, 8:0] and sequences {2, 3} is another.

maximal repeats within a single path; we can mark and ignore all internal suffix tree nodes that contain only a single haplotype path sequence in linear time using a standard method [9]. Thus, a simple procedure for locating pangenome haplotype blocks is as follows:

1. Build the string $\mathbb{P} = p_1\$_1p_2\$_2 \dots p_k\$_k$, where each $\$_i$ is a distinct character not used in the $p_i$ strings.
2. Build a suffix tree on $\mathbb{P}$.
3. Use the suffix tree to find all maximal repeats $(K, S)$ in $\mathbb{P}$. The SNP path and the set of sequences $K$ are represented implicitly by the suffix tree node.

Building a suffix tree can be done in $O(nk)$ time and space [5], and, as noted above, finding all maximal repeats in the suffix tree is also linear time. Thus, each step of the procedure takes linear time and space.

Since the MPPHBs correspond to internal nodes in the suffix tree on $\mathbb{P}$, we can give a bound on the number of MPPHB in $P$.

**Lemma 1.** *Given a set of $k$ pangenome paths $P$ with maximum length $n$, there are at most $(n+1)k$ MPPHBs in $P$.*

*Proof.* As argued above, every MPPHB in $P$ corresponds to a maximal repeat in $\mathbb{P}$. Because each path in $P$ contains no more than $n$ nodes, $|\mathbb{P}| \leq (n+1)k$. Then, because the maximal repeats of a string are the internal nodes in the suffix tree of that string [7, Theorem 7.12.1], there are at most $(n+1)k$ maximal repeats in $\mathbb{P}$, and thus at most $(n+1)k$ MPPHBs in $P$.

## 6    Experimental Results

We tested our method for finding MPPHBs using a moderately-sized pangenomic yeast data set. Yeast is a well-studied model system with a genome size of approximately 12 Mb. We created a yeast data set using assemblies from 10 yeast strains from the *Saccharomyces* Genome Database[1] used in either wine or bread-making. To investigate the maximal perfect pangenome haplotype blocks present in the data set, we construct a compressed De Bruijn graph for $k \in \{25, 100, 1000\}$ using the `cdbg` package [2] and extract SNPs from each using the method described in Sect. 3. Each yeast sequence then corresponds to a path through the SNP graph $p_i$; that is, a sequence of pangenome SNP calls. Then, as in Sect. 5, we find maximal repeats in the string $p_1\$_1p_2\$_2 \dots p_k\$_k$ in order to find MPPHBs. We use `repeat-match` from MUMmer 4.0 [8] to compute maximal repeats and identify all maximal pangenomic haplotype blocks using these reported repeats.

Compressed De Bruijn graph and SNP graph generation takes a few minutes on a moderate workstation[2] for this data set. In order to find maximal repeats

---

[1] http://www.yeastgenome.org The strains used were AWRI796 (Wine), BC187 (Wine), CLIB215 (Bakery), CLIB324 (Bakery), DBVPG6044 (Wine), L1528 (Wine), LalvinQA23 (Wine), Red Star (Bakery), VL3 (Wine), YS9 (Bakery).

[2] An 8-core 3.40 GHz Intel i7 CPU with 16 Gb of RAM.

using MUMmer, we encode SNP nodes using 19 alphabet characters. When running `repeat-match`, we use the `-f` flag to find forward repeats only and the `-n` flag to return only encoded repeats long enough to represent full SNP nodes (in our case, 19 characters). For all $k$ values, `repeat-match` took at most a few seconds to run. We then use a simple Python script to decode the output back to SNP labels and process it into haplotype blocks. For $k = 25$ and $k = 100$, this takes a few minutes; for the other two values tested, it takes a few seconds or less.

Table 1 shows the number of SNPs found in each experiment, as well the number of haplotype blocks found and their average number of sequences and SNP path length. When $k = 1000$ fewer SNPs are found since there are fewer bubbles in the cDB graph and the blocks are smaller in size. As the number of bubbles in the cDB graph increases, more blocks are found. We leave a more thorough investigation of the relationship between $k$, the number of bubbles, and the number of blocks to future work.

**Table 1.** Summary statistics for different $k$ values. Decreasing $k$ from 1000 to to 25 results in a larger SNP graph and more and bigger blocks found.

| k | # SNPs | # blocks | avg. $|K|$ | avg. $|S|$ |
|---|---|---|---|---|
| 1000 | 1,985 | 146 | 2.12 | 1.47 |
| 500 | 4,759 | 1,458 | 2.46 | 1.68 |
| 100 | 38,489 | 39,036 | 3.46 | 3.61 |
| 25 | 117,792 | 79,154 | 3.41 | 4.24 |

We compare the distributions of these data for $k = 500$ and $k = 100$ in Fig. 4.

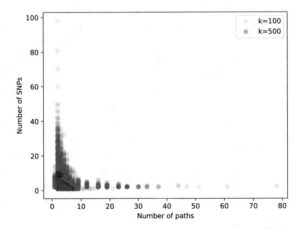

**Fig. 4.** Scatterplot showing each distribution of maximal perfect pangenome haplotype sizes for $k = 100$ and $k = 500$.

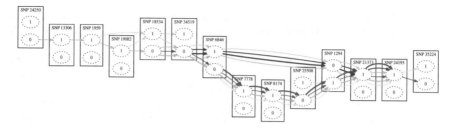

**Fig. 5.** Sample haplotype block paths from a pangenomic data set comprised of 10 yeast genomes. Each colored path represents a haplotype block and the line thickness is proportional to the number of sequences in the block. SNPs 7778, 8174 and 25508 represent an introgressed region. (Color figure online)

In Fig. 5 we show a plot of several of the maximal haplotype blocks found in the $k = 100$ graph. The graph shows an introgressed region of SNPs that occurs in approximately half of the sequences that traverse the region shown.

## 7 Conclusion

In this work, we define the maximal perfect pangenome haplotype block problem and give a linear time method to solve it. Single-reference haplotype blocks can be used to compute a selection coefficient measuring the relative fitness of two genetic variants in a population; a natural next step in the pangenome case is to precisely define a pangenomic selection coefficient based on MPPHBs, or to explore other applications of MPPHBs in population genetics.

We note that the positional Burrows-Wheeler Transform approach from [1] cannot be directly adapted for pangenome haplotype blocks since the SNP graph is not generally linear and paths may skip SNPs or contain cycles, etc. However, we are interested in extending both the pangenome and single-reference maximal perfect haplotype block problem to include inputs with SNPs that are not called, in order to include genomes with low coverage in some regions.

**Acknowledgements.** Support provided by US National Science Foundation grants DBI-1759522 and DBI-1661530. We thank the anonymous reviewers for their thoughtful feedback and questions.

## References

1. Alanko, J., Bannai, H., Cazaux, B., Peterlongo, P., Stoye, J.: Finding all maximal perfect haplotype blocks in linear time. In: 19th International Workshop on Algorithms in Bioinformatics, WABI 2019. Schloss Dagstuhl-Leibniz-Zentrum fuer Informatik (2019)
2. Beller, T., Ohlebusch, E.: A representation of a compressed de Bruijn graph for pan-genome analysis that enables search. Algorithms Mol. Biol. **11**(1), 20 (2016)

3. Chen, H., Hey, J., Slatkin, M.: A hidden markov model for investigating recent positive selection through haplotype structure. Theoret. Popul. Biol. **99**, 18–30 (2015)
4. Cunha, L., Diekmann, Y., Kowada, L., Stoye, J.: Identifying maximal perfect haplotype blocks. In: Alves, R. (ed.) BSB 2018. LNCS, vol. 11228, pp. 26–37. Springer, Cham (2018). https://doi.org/10.1007/978-3-030-01722-4_3
5. Farach, M.: Optimal suffix tree construction with large alphabets. In: Proceedings 38th Annual Symposium on Foundations of Computer Science, pp. 137–143. IEEE (1997)
6. Gillespie, J.H.: Population Genetics: a Concise Guide. JHU Press, Baltimore (2004)
7. Gusfield, D.: Algorithms on Strings, Trees, and Sequences: Computer Science and Computational Biology. Cambridge University Press, Cambridge (1997)
8. Marçais, G., Delcher, A.L., Phillippy, A.M., Coston, R., Salzberg, S.L., Zimin, A.: MUMmer4: a fast and versatile genome alignment system. PLoS Comput. Biol. **14**(1), e1005944 (2018)
9. Sung, W.K.: Algorithms in Bioinformatics: A Practical Introduction. CRC Press, Boca Raton (2009)
10. Tettelin, H., et al.: Genome analysis of multiple pathogenic isolates of streptococcus agalactiae: implications for the microbial "pan-genome". Proc. Natl. Acad. Sci. **102**(39), 13950–13955 (2005)

# Gaps and Runs in Syntenic Alignments

Zhe Yu, Chunfang Zheng, and David Sankoff[(✉)] [iD]

University of Ottawa, Ottawa, Canada
{zyu096,czhen033,sankoff}@uottawa.ca

**Abstract.** Gene loss is the obverse of novel gene acquisition by a genome through a variety of evolutionary processes. It serves a number of functional and structural roles, compensating for the energy and material costs of gene complement expansion.

A type of gene loss widespread in the lineages of plant genomes is "fractionation" after whole genome doubling or tripling, where one of a pair or triplet of paralogous genes in parallel syntenic contexts is discarded.

The detailed syntenic mechanisms of gene loss, especially in fractionation, remain controversial.

We focus on the the frequency distribution of gap lengths (number of deleted genes – not nucleotides) within syntenic blocks calculated during the comparison of chromosomes from two genomes. We mathematically characterize s simple model in some detail and show how it is an adequate description neither of the *Coffea arabica* subgenomes nor its two progenitor genomes.

We find that a mixture of two models, a random, one-gene-at-a-time, model and a geometric-length distributed excision for removing a variable number of genes, fits well.

**Keywords:** Gene loss · Tetraploidy · Fractionation · Plant genomes · Coffee · Run length

## 1 Introduction

The evolutionary process of gene loss, through DNA excision, pseudogenization or other mechanism, is the obverse of novel gene acquisition by a genome through processes such as tandem duplication, gene family expansion, whole genome doubling, neo- and subfunctionalization and horizontal transfer. Loss serves a number of functional and structural roles, mainly compensating for the energetic, material and structural costs of gene complement expansion.

A type of gene loss widespread in the lineages of plant genomes, and also occurring in a few yeast, fish and amphibian genomes, is "fractionation" after whole genome doubling or tripling, where one of a pair or triplet of paralogous genes in parallel syntenic contexts is discarded.

Quantitative studies have focused on many aspects of gene loss. In this paper, we study the evolutionary history of the allotetraploid *Coffea arabica* (CA) and

© Springer Nature Switzerland AG 2020
C. Martín-Vide et al. (Eds.): AlCoB 2020, LNBI 12099, pp. 49–60, 2020.
https://doi.org/10.1007/978-3-030-42266-0_5

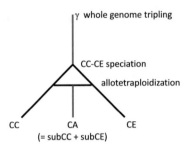

**Fig. 1.** *Coffea* phylogeny. Fractionation operates in lineages coloured red. (Color figure online)

its two diploid progenitors, *Coffea canephora* (CC) and *Coffea eugenioides* (CE), annotated genome assemblies being provided by the Arabica Coffee Genome Consortium [1]. This history is summarized in Fig. 1. We survey gene loss in three periods. These are

- loss from the ancestral lineage leading from the γ whole genome tripling event [2] 120 million years ago, due at least partly to fractionation,
- independent losses from the CC and CE genomes after speciation (but before allotetraploidization) around 10 million years ago [3], and
- loss from the CC and CE, and the subCC and subCE subgenomes of CA, following the allotetraploidization event. Loss from the two subgenomes (namely those chromosomes in CA deriving from CC and those deriving from CE) can be attributed to fractionation.

We first study the distribution of gene pair similarities derived from the comparison of the four genomes and subgenomes. This will serve to confirm the validating parallels between CC and CE evolution, and between subCC and subCE evolution.

We then introduce our main analytical construct, the frequency distribution of gap lengths within syntenic blocks calculated during the comparison of chromosomes from two genomes or subgenomes. In the simplest model, proposed over ten years ago [4–6], at each step a random gene pair is selected to lose one member. In a new version of this model that takes into account chromosome length, we develop an exact recurrence to calculate the expected number of gaps of each length after a given number of steps. We then provide evidence from the *Coffea* data that demonstrates a systematic departure from this model.

In a competing class of models [7,8], gene loss is effected by excision of a variable length fragment of a chromosome, often formulated in terms of a gamma distribution. In the *Coffea* data, there are far too many single-gene deletions for this solution, but a mixture of the two models, where the gamma is actually a single-parameter geometric distribution, fits well.

## 2    Methods

Our research is based on the homologous gene pairs in syntenic context as produced from the data on pairs of genomes by the SYNMAP procedure on the CoGE platform [9,10]. At a general level, we used the "peaks" method [11] for the three events that generate duplicate genomes in the evolution of CA: gamma hexaploidization, CC/CE speciation and CA tetraploidization (which is effectively a speciation of CC/CA-subCC and of CE/CA-subCE). In this method, the local modal values (peaks) of the distribution of the entire set of homologous gene pairs, as calculated by the R function geom_density, are estimates of the time of the event. We could also have used EMMIX [12] or other mixture of distributions methods to carry this out.

This allowed us to study the evolution of paralogous and orthologous synteny blocks. We considered only genes within the region of the blocks, including gene pairs and singleton genes in each genome that have lost their counterpart in the other genome due to fractionation or other gene loss. We used all four genomes, CC, CE, CA-subCC (denoted just subCC) and CA-subCE (denoted just subCE), producing six comparisons of pairs, and four self-comparisons. We did not look at the whole CA assembly, just the large majority that was successfully separated into the subgenomes.

We studied a number of statistics on the gaps between adjacent pairs of duplicate genes within synteny blocks, the innovative focus of this work, and here report on one of them, the size of gaps between two adjacent duplicate pairs on genes in a block, from 0 (no gap) to a maximum of 10 on either one of the genomes. We make certain operational definitions to allow us to analyze evolution coherently across all evolutionary eras. For example, if we encounter more than 10 genes in a gap on one genome between two adjacent gene pairs, we break up the synteny block into two at that point. This is justified by the regular decrease in frequency in gap sizes from 0, 1, 2, until there are almost none of size 8, 9, or 10, except between neighbouring synteny blocks, which can be separated by large numbers of unpaired genes in either of both of the genomes. We want to study the nature of the distribution of gap size due to fractionation or gene deletion, and this avoids biasing estimates by inclusion of gaps produced by mechanisms other than fractionation. Thus, we use the default parameters of SYNMAP, except for the maximum number of non-duplicate genes interrupting any neighbouring gene pairs, which we set at 10.

## 3    Results

### 3.1    The Sequence of Evolutionary Events

Some 28,800, 33,500 and 56,700 genes were identified in the annotations of CC, CE and CA, respectively, while the subCC and subCE subgenomes identified in CA contained 24,700 and 25,800 genes respectively. Amalgamating the gene pairs in all SYNMAP comparisons produces the CIRCOS plot in Fig. 2.

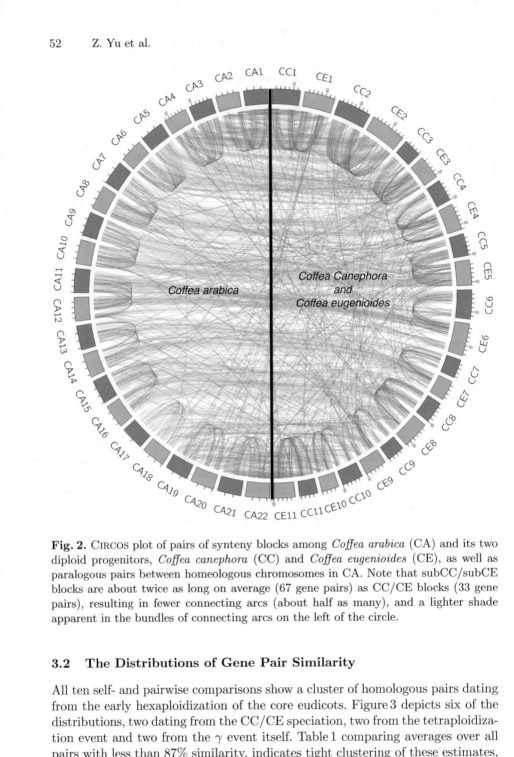

**Fig. 2.** CIRCOS plot of pairs of synteny blocks among *Coffea arabica* (CA) and its two diploid progenitors, *Coffea canephora* (CC) and *Coffea eugenioides* (CE), as well as paralogous pairs between homeologous chromosomes in CA. Note that subCC/subCE blocks are about twice as long on average (67 gene pairs) as CC/CE blocks (33 gene pairs), resulting in fewer connecting arcs (about half as many), and a lighter shade apparent in the bundles of connecting arcs on the left of the circle.

## 3.2   The Distributions of Gene Pair Similarity

All ten self- and pairwise comparisons show a cluster of homologous pairs dating from the early hexaploidization of the core eudicots. Figure 3 depicts six of the distributions, two dating from the CC/CE speciation, two from the tetraploidization event and two from the $\gamma$ event itself. Table 1 comparing averages over all pairs with less than 87% similarity, indicates tight clustering of these estimates, in terms of peak gene similarity (over CDS regions).

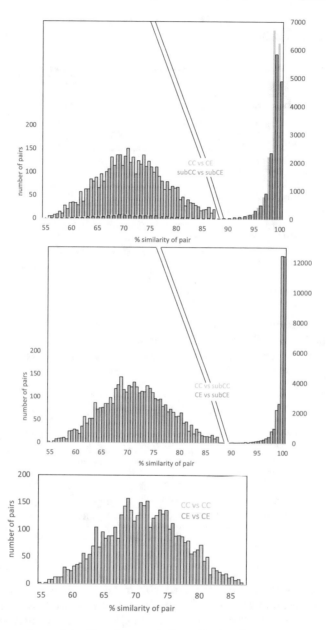

**Fig. 3.** Gene pairs originating in speciation (top). Gene pairs originating in tetraploidization (middle). Gene pairs originating in $\gamma$ event (bottom)

**Table 1.** Locating the $\gamma$ hexaploidy in all comparisons; peak of distribution of pairs with less than 87% similarity.

| Comparison | Peak of similarity (%) | # of pairs |
|---|---|---|
| CC vs CE | 73.7 | 2069 |
| subCC vs CE | 67.8 | 1860 |
| subCE vs CC | 68.0 | 2105 |
| subCC vs subCE | 67.9 | 1922 |
| CC vs CC | 71.5 | 1163 |
| CE vs CE | 70.4 | 1056 |
| subCC vs subCC | 68.2 | 938 |
| subCE vs subCE | 70.1 | 1043 |
| subCC vs CC | 68.2 | 1949 |
| subCE vs CE | 67.8 | 1925 |
| Mean ± S.D. | 69.4 ± 2.0 | 1603 ± 484 |

The speciation of CC and CE generates orthologous gene pairs visible in the CC (or subCC) vs CE (or subCE) comparisons, as can be seen on the right hand side cof the top panel in Fig. 3. Table 2 presents the peak similarity for these comparisons.

**Table 2.** Locating the CC-CE speciation

| Comparison | Similarity (%) | # of pairs |
|---|---|---|
| CC vs CE | 99.12 | 17,066 |
| subCC vs CE | 99.08 | 15,985 |
| subCE vs CC | 99.11 | 16,014 |
| subCC vs subCE | 99.05 | 15,318 |
| Mean ± S.D | 99.09 ± 0.033 | 16,096 ± 626 |

The CA tetraploidization event, which for our purposes consists of the synchronous speciation of CC/subCC and CE/subCE, is visible as a peak of gene pairs in the CC vs subCC and the CE vs subCE comparisons on the right hand side of the middle panel in Fig. 3. These peaks are listed in Table 3.

The current best estimates of $\gamma$ and CC/CE *Coffea* speciation are of the order of 120 My and 10 My [3], while the CA tetraploidization is thought to be less than 1 My old. The similarity measures does not correspond well to this timeline. The tetraploidy seems to be 15–20% of the speciation age.

**Table 3.** Locating the tetraploidization event

| Comparison | Similarity (%) | # of pairs |
|---|---|---|
| CC vs subCC | 99.81 | 16,487 |
| CE vs subCE | 99.87 | 17,196 |
| Mean ± S.D. | 99.84 ± 0.043 | 16,842 ± 501 |

## 4  One-at-a-Time Model

Consider the following "fractionation" process. We have an array of $n$ 1's. At the first step, and every subsequent step, we pick a 1 at random and transform it to 0. We stop after a given number of steps $t \leq n$.

We prove a recurrence for $M(t, x)$, the expected number of runs of 1's (more precisely, maximal runs) of length $x$ at time $t$.

**Proposition 1**

$$M(0, n) = 1$$
$$M(0, x) = 0, \quad \text{for } x \neq n \tag{1}$$

*Thereafter, for $1 \leq t \leq n - 1$ and $1 \leq x < n - t + 1$*

$$M(t, x) = M(t - 1, x) - \frac{x M(t - 1, x) - 2 \sum_{i > x}^{n} M(t - 1, i)}{n - t + 1}, \tag{2}$$

*Proof.* The initial values of the process at $t = 0$ are fixed by definition, and so then are their averages $M(0, x)$.

For each $t > 0$, in randomly changing one of the $n - t + 1$ remaining 1's in the array to 0, there are two mutually exclusive possibilities. An existing run of length $x$ can be destroyed, for some $x \geq 1$, which can happen $x M(t - 1, x)$ ways. Alternatively a run of length $x$ can be created. This can occur in exactly two ways in breaking up any remaining run of length greater than $x$.

The average change is obtained through division by the total number of cases $n - t + 1$.                                                                                  □

There is a symmetry in the fractionation process, in that the evolution of the number of 1's, and the probabilistic structure governing the distribution of run sizes, starting from time $t = 0$, is identical to the evolution of the number of 0's, and the probabilistic structure governing the distribution of gap sizes, starting from time $t = n$.

To illustrate the the evolution of run lengths, Fig. 4 shows how longer runs only survive at the beginning of the process, and how the number of shorter runs increases until they too are lost to fractionation. Of interest is the case of run length 2, where the symmetry of gaps and runs is clearest.

This process bears much resemblance to the theory of runs [13] in random binary sequences. Given $n$ Bernoulli trials with a probability of success $p = t/n$,

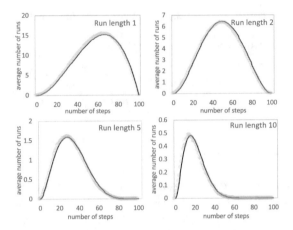

**Fig. 4.** Evolution of number of runs of 1's of various sizes as the number of steps $t$ increases. Solid line: recurrence. Light blue background line: average of 1000 simulations. Genome length $n = 100$ (Color figure online)

the expected number of successes is $t$, and the expected number of runs of length $x$ is $M(t, x)$. However, the variance of the number of successes is non-negligible, whereas it is zero for our process, and the variance of the number of runs of a given length is also greater than our process. Thus our interest in the fractionation process, where the probability of success at each position depends on the total number of successes already achieved.

When we compare the behaviour of our "one-at-a-time" model with the gaps in the *Coffea* data in Fig. 5 however, it is clear that the model is inadequate to account for both the simultaneous steep drop-off from gaps of size 1 and the presence of significant number of long gaps.

## 4.1   The Combined Model

To remedy the poor fit of the one-at-a-time model, we combine it with a gamma distribution component. Whereas the one-at-a-time model involves a single fixed parameter, chromosome size $n$, a gamma component adds a shape and a scale parameter, as well as weight parameter to apportion the two components. Fortunately, the optimal gamma component turned out to be a simple geometric distribution, with only one parameter.

To estimate the $n$, the geometric parameter $\lambda$, and the proportion of steps $\theta$ allocated to the one-at-a-time model, we compared data on runs of 1's and runs of 0's, from the both the speciation event and the tetraploidization event, all taken together. We optimized in terms of a chi-square criterion, when running 50 simulations based on a range of values of $n$, $\lambda$ and $\theta$. (This was after finding that the two parameters of a general gamma distribution did not substantially improve the fit compared to a geometric distribution.) Of importance, however,

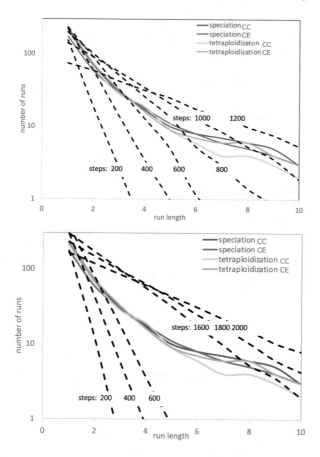

**Fig. 5.** Inability of one-at-a-time model to fit *Coffea* data on runs of zeros (gaps), with chromosome size $n = 1600$ (top) or $n = 2800$ (bottom), and at various time intervals (steps)

was that we allowed different numbers of steps in the speciation and tetraploid simulations. The values were 705 for speciation and 590 for tetraploidization, which is coherent with the historical ordering of these two events, and with the mean similarities in Tables 2 and 3. The optimal values were $\theta = 0.7$, $n = 2800$ and $\lambda = 2/7$.

Our model breaks down when we use it to simulate fractionation after $\gamma$, as can be seen in the bottom panel of Fig. 6. The simulations suggest there should remain no long runs of 1's, but this is likely due to the inability to detect sufficiently long synteny blocks after extensive fractionation, and possibly some tendency for some regions of neighbouring genes to resist fractionation.

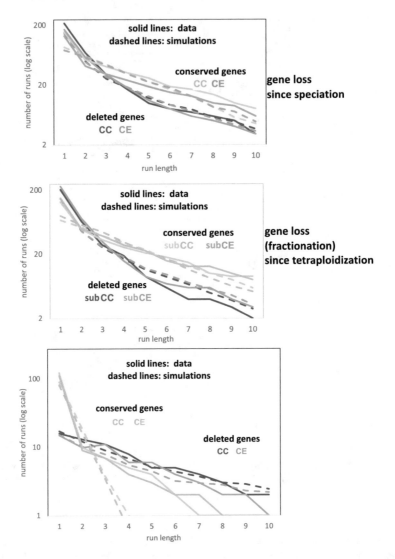

**Fig. 6.** Comparisons of the distributions of run sizes (0's and 1's) with simulations of combined model in the speciation data (top panel), tetraploidization data (middle panel) and $\gamma$ data (bottom panel).

## 5    Conclusions

It can be noted that in all of our comparisons, there has been a symmetry between CC and CE, and between subCC and subCE. If $\gamma$ fractionation rates or evolutionary divergence rates of CC and CE or subgenome dominance play a role, their effects must be relatively small.

We have found several indications that the time span since tetraploidization, is almost as long as the period since speciation.

We have investigated the one-at-a-time model in some detail, but it is clearly inadequate to explain the gene loss data, which is surprisingly parallel between post-speciation loss and fractionation. Adding a geometric component, however, allows the model to fit the data quite well.

The distribution of gap sizes in synteny blocks generated by all evolutionary events confirms that gene loss, by fractionation or otherwise, proceeds largely by the loss of one gene at a time, with $\theta = 0.7$ and further loss from the geometric component.

**Acknowledgments.** Research supported in part by grants from the Natural Sciences and Engineering Research Council of Canada. DS holds the Canada Research Chair in Mathematical Genomics.

# References

1. De Kochko, A., Crouzillat, D.: Arabica coffee genome consortium: Aims and goals of the Arabica Coffee Genome Consortium (ACGC). In: 12th Solanaceae Conference (2015)
2. Jaillon, O., Aury, J.M., Noel, B., Policriti, A., Clepet, C., et al.: The grapevine genome sequence suggests ancestral hexaploidization in major angiosperm phyla. Nature **449**, 463–467 (2007)
3. Hamon, P., Grover, C.E., et al.: Genotyping-by-sequencing provides the first well-resolved phylogeny for coffee (Coffea) and insights into the evolution of caffeine content in its species: GBS coffee phylogeny and the evolution of caffeine content. Mol. Phylogenet. Evol. **109**, 351–361 (2017)
4. van Hoek, M.J., Hogeweg, P.: The role of mutational dynamics in genome shrinkage. Mol. Biol. Evol. **24**, 2485–2494 (2007)
5. Byrnes, J.K., Morris, G.P., Li, W.H.: Reorganization of adjacent gene relationships in yeast genomes by whole-genome duplication and gene deletion. Mol. Biol. Evol. **23**, 1136–1143 (2006)
6. Zheng, C., Wall, P.K., Leebens-Mack, J., dePamphilis, C., Albert, V.A., Sankoff, D.: Gene loss under neighbourhood selection following whole genome duplication and the reconstruction of the ancestral *Populus* diploid. J. Bioinform. Comput. Biol. **7**, 499–520 (2009)
7. Sankoff, D., Zheng, C., Wang, B., Fernando Buen Abad Najar, C.: Structural vs. functional mechanisms of duplicate gene loss following whole genome doubling. BMC Genomics **15** (2015). https://doi.org/10.1109/ICCABS.2014.6863915
8. Yu, Z.N., Sankoff, D.: A continuous analog of run length distributions reflecting accumulated fractionation events. BMC Bioinform. **17**(suppl 14), 412 (2016)
9. Lyons, E., Freeling, M.: How to usefully compare homologous plant genes and chromosomes as DNA sequences. Plant J. **53**, 661–673 (2008)
10. Lyons, E., et al.: Finding and comparing syntenic regions among *Arabidopsis* and the outgroups papaya, poplar and grape: CoGe with rosids. Plant Physiol. **148**, 1772–1781 (2008)

11. Sankoff, D., Zheng, C., Zhang, Y., Meidanis, J., Lyons, E., Tang, H.: Models for similarity distributions of syntenic homologs and applications to phylogenomics. IEEE/ACM Trans. Comput. Biol. Bioinform. **16**, 727–737 (2019)
12. McLachlan, G.J., Peel, D., Basford, K.E., Adams, P.: The EMMIX software for the fitting of mixtures of normal and t-components. J. Stat. Softw. **4**, 1–14 (1999)
13. Weisstein, E.: Run. MathWorld-A Wolfram Web Resource. http://mathworld. wolfram.com/. Accessed 20 Aug 2019

# Phylogenetics

# Comparing Integer Linear Programming to SAT-Solving for Hard Problems in Computational and Systems Biology

Hannah Brown[1] , Lei Zuo[1,2] , and Dan Gusfield[1]([✉])

[1] Department of Computer Science, University of California,
Davis 1 Shields Avenue, Davis, CA 95616, USA
{hsbrown,leizuo,dmgusfield}@ucdavis.edu
[2] Department of Computer Science, The University of Hong Kong,
Pok Fu Lam, Hong Kong, China

**Abstract.** It is useful to have general-purpose solution methods that can be applied to a wide range of problems, rather than relying on the development of clever, intricate algorithms for each specific problem. Integer Linear Programming is the most widely-used such general-purpose solution method. It is successful in a wide range of problems. However, there are some problems in computational biology where integer linear programming has had only limited success. In this paper, we explore an alternate, general-purpose solution method: SAT-solving, i.e., constructing Boolean formulas in conjunctive normal form (CNF) that encode a problem instance, and using a SAT-solver to determine if the CNF formula is satisfiable or not. In three hard problems examined, we were very surprised to find the SAT-solving approach was dramatically better than the ILP approach in two problems; and a little slower, but more robust, in the third problem. We also re-examined and confirmed an earlier result on a fourth problem, using current ILP and SAT-solvers. These results should encourage further efforts to exploit SAT-solving in computational biology.

**Keywords:** Integer programming · SAT-solving · Computational biology

## 1 Introduction

Integer (Linear) Programming, abbreviated "ILP", is a versatile modeling and optimization technique that has been increasingly used in *computational and systems biology* in *non-traditional* ways. ILP is often (but not always) very effective in solving *instances* of hard computational problems on *realistic* biological

Thanks to NSF for supporting this research under the grant 1528234. Support for H.B. was from NSF Research Experiences for Undergraduates grant 1528234. Thanks also to Chase Maguire for his contributions at the start of this project, and thanks to the very helpful and in-depth conference reviews we received.

C. Martín-Vide et al. (Eds.): AlCoB 2020, LNBI 12099, pp. 63–76, 2020.
https://doi.org/10.1007/978-3-030-42266-0_6

data of current importance. See [8] for an in-depth discussion of ILP in computational and systems biology, illustrating many successes (and some failures) of the ILP approach in computational and systems biology.

Despite the many successes of ILP in computational and systems biology, in the testing of ILP formulations during the writing of [8], we observed a few computational biology problems where the ILP approach was only moderately effective, or was almost completely ineffective. For such hard problems, it is of interest to try different well-developed, general, computational techniques. The most well-developed such technique involves formulating a computational problem as a Boolean formula in Conjunctive Normal Form (CNF), and using a *SAT-solver* to determine whether that CNF formula is satisfiable. We wrote programs to create CNF-formulations for four hard computational biology problems, and tested these using the widely-used SAT-solver, *pLingeling* [3]. We wanted to see if this SAT approach could solve problem instances that were difficult for the ILP approach, using the highly-regarded ILP solver created by *Gurobi Optimization*. We then did some additional testing using another highly-regarded SAT-solver, *Glucose-Syrup*, to check that our results were not just artifacts of using pLingeling.

Previous comparisons of ILP and SAT found problems (not from computational biology) where the ILP approach was highly efficient but the SAT approach was ineffective [11]. In contrast, two other comparisons (for problems in computational biology) showed that the SAT approach was much faster than the ILP approach [15–17].

*In this Paper.* We report on the development of SAT formulations, and in-depth empirical comparisons of ILP and SAT-solving for *three* hard problems, where prior ILP approaches were much less effective than desired: *protein folding* under the *HP model*; *transforming gene orders by reversals*; and computing the *History Bound* on the number of recombinations needed to generate a given set of SNP sequences. We also re-implemented and confirmed an earlier published result on the use of SAT-solving for the problem of *haplotyping by pure parsimony*. All of the software we developed is freely available online, as are the SAT-solvers we used, *pLingeling* (mostly) and *Glucose-Syrup*. The ILP-solver we used, from Gurobi Optimization, is available with a free academic license.

*Results.* In the protein folding problem, the ILP approach was seen to be generally faster than the SAT approach, with some notable exceptions. But in the other two newly-examined problems (reversals and History Bound), the SAT approach was substantially superior to the ILP approach in terms of solution times, and in the avoidance of extreme behaviors that the ILP approach sometimes exhibits. Also, our reexamination of pure-parsimony haplotyping confirmed the results in [15,16], and found extreme cases where SAT solved in a practical amount of time, but the ILP approach made little progress towards a solution.

We examine each of these hard problems in the next sections. We start with the protein folding problem, presenting the full logic and details of the CNF formulas in order to illustrate the SAT approach. Space limits the full details of the CNF formulas in discussing the other three problems. These will be detailed in an expanded paper to be written.

## 2   Protein Folding via the HP Model

Determining the three-dimensional structure of proteins, or learning some parts of the structure, are critical tasks in biochemistry, biophysics, computational biology, and systems biology. For computational effectiveness, an attractive simple model of globular protein folding, the *HP model*, was proposed by Dill [5] in 1995, and has been extensively explored since then. The model simplifies the twenty standard amino acids by dividing them into *two* groups: the *hydrophobic* (H) and the *hydrophilic* (P); that is, water-fearful and water-loving. Hence a protein sequence, based on an alphabet of size twenty, is reduced to a binary sequence (H or P). For the biological motivation for this model, see [8].

*The HP Prototein Model and Folding Problem in 2-D.* A *prototein*[1] is a binary sequence that we embed on a *two-dimensional grid*, where each 1 encodes an H and each 0 encodes a P. A *legal embedding* of a prototein on the grid must satisfy the following rules:

1. Each character in the sequence gets assigned to some point on the grid.
2. No character in the sequence gets assigned to more than one point on the grid.
3. Each point of the grid gets assigned *at most* one character in the sequence.
4. Two adjacent characters in the sequence must be placed on two points that are *neighbors* on the grid in either the horizontal or vertical direction, but not both.

These four rules mean that the sequence must be embedded into the grid as a *self-avoiding walk*, without deforming the sequence.

In an embedding of the sequence, we say that a *potential contact* exists between neighboring points $j$ and $m$ on the grid $G$ if and only if a 1 is assigned to both points $j$ and $m$ of $G$. A potential contact $(j, m)$ is a (real) *contact* if the 1s that are assigned to points $j$ and $m$ are *not* adjacent in the sequence. The central assumption in the HP model is that the most stable fold is one that maximizes the number of contacts (H-H bonds). This leads to:

**The 2-D Prototein Problem.** Given a binary sequence $S$, find a legal embedding of the sequence in the 2-D grid, $G$, to maximize the number of contacts.

We define the *offset* as the number of adjacent positions in sequence $S$ that both have character 1. Clearly, the number of contacts in an embedding is the number of potential contacts minus the offset. Since we can easily determine the offset given $S$, the problem of maximizing the number of contacts can be solved by maximizing the number of potential contacts, and that is the approach we take. The 2-D Prototein Problem has been naturally generalized to a 3-D grid, which better models real protein folding.

---

[1] Yes, the word is "prototein" not "protein".

*ILP and SAT Approaches to the Prototein Problem.* Previously [8,21], ILP formulations were developed to solve the 2-D and 3-D Prototein Problems, and empirical results were reported in [8,21]. The ILP formulations directly solve the optimization version of the problem, while the SAT formulation can only implement a *decision* version. The maximum number of contacts is then obtained by solving a series of such SAT decision problems, changing the target each time.

*The Logic for the SAT Formulation.* Given a target number of contacts, the SAT formulation must ensure that binary sequence $S$ is legally embedded into grid $G$; have logic to represent what is and isn't a potential contact; and have logic to count the number of potential contacts in the embedding. The resulting CNF formula will be satisfied if and only if there is a legal embedding where the number of potential contacts meets or exceeds the target number.

The points on the grid we use are numbered from 1 to $w^2$ for the 2-D version of the problem, and from 1 to $w^3$ in the 3-D version. So, each point on the grid is indexed by a single number. Clearly, $w$ can be set to $n$, but choosing a smaller diameter results in much faster computation. We will discuss this later.

*Ensuring Legal Embeddings.* To represent the conditions for a legal embedding, we use variables $X_{i,j}$, where $i$ is a character-position in $S$ and $j$ is a point in $G$. If variable $X_{i,j}$ is set true, then the character at position $i$ in $S$ is assigned to point $j$ in $G$. There are four conditions that ensure a legal embedding (in both 2-D and 3-D) of a binary sequence $S$:

1. Every character $i \in S$ must be assigned to some point $j \in G$. The CNF formulas for this condition are shown in Eq. 1.

$$\bigvee_{1 \leq j \leq |G|} X_{i,j} \tag{1}$$

2. No character-position $i \in S$ can be assigned to more than one point $j \in G$. The CNF formulas for this condition are shown in Eq. 2.

$$\bigwedge_{\substack{1 \leq j,j' \leq |G| \\ j \neq j'}} (\overline{X_{i,j}} \vee \overline{X_{i,j'}}) \tag{2}$$

3. No point $j \in G$ can have more than one character-position $i \in S$ assigned to it. The CNF formulas for this condition are similar to those for condition 2, varying over $i$ and $i'$ rather than $j$ and $j'$.

4. Every adjacent pair of character-positions $i, i+1 \in S$ must be placed on adjacent points in $G$ (diagonals are not adjacent). So, for each position $i \in S$ and point $j \in G$, we need the CNF clause shown in Eq. 3.

$$\overline{X_{i,j}} \bigvee_{1 \leq j' \leq |G| \text{ j and j' are neighbors}} X_{i+1,j'} \tag{3}$$

*Grid Size.* To limit the number of variables, we set the grid diameter $w$ to $1 + \lfloor \frac{n}{4} \rfloor$ for all values of n $\geq$ 12, and to $n$ otherwise. Likewise, in the 3-D version, $w$ is set to $2 + \lfloor \frac{n}{8} \rfloor$ for all values of $n \geq 20$ and to $2 + \lfloor \frac{n}{4} \rfloor$ otherwise.

*Identifying Potential Contacts.* There are two conditions for a potential contact:

1. A potential contact involves point $j \in G$ if and only if a character 1 is assigned to point $j$. The variable $T_j$ is set true if and only if this condition holds. The CNF formulas for this are omitted for lack of space.
2. A potential contact exists between points $j$ and $m$ on the grid if and only if a 1 is assigned to both points $j$ and to a neighboring point $m$ in G. The variable $C_{j,m}$ will be set to true if both $T_j$ and $T_m$ are set true. The CNF formulas for this are shown in Eq. 4.

$$(\overline{C_{j,m}} \vee T_j) \wedge (\overline{C_{j,m}} \vee T_m) \wedge (C_{j,m} \vee \overline{T_j} \vee \overline{T_m}),$$
$$\forall \text{ neighboring positions } j, \ m, 1 \leq j, \ m \leq |G| \tag{4}$$

*Counting Potential Contacts.* It is not obvious how to use CNF formulas to *count* the number of potential contacts in an embedding, i.e., the number of $C$ variables set true. To implement counting we use a non-trivial method detailed in [11] and originating in [2], which, given a goal number of potential contacts, $m$, counts the number of $C$ variables that are *not* set true and tests whether it is less than or equal to $r$, where $r = 2|G| - m$ for the 2-D version and $r = 3|G| - m$ for the 3-D version. The CNF formulas for this are shown in Eq. 5.

$$(\overline{b_i}^{2k} \vee \overline{b_j}^{2k+1} \vee b_{i+j}^k),$$
$$0 \leq i \leq t_{2k}, \ 0 \leq j \leq t_{2k+1}, \ 1 \leq i+j \leq t_k + 1, 1 < k < |G| \tag{5}$$
$$(\overline{b_i}^2 \vee \overline{b_j}^3), \ 0 \leq i \leq t_2, \ 0 \leq j \leq t_3, \ i+j = r+1$$

The $b$ variables exist for all internal nodes in a binary tree with $c = \lceil log_2(r) \rceil$ levels. Each internal node is assigned $t_k = min(r, 2^{c-d})$ $b$ variables, where $d$ is the depth of node $k$. The leaves are set to $\overline{C}$, and $b_i^k$ being set true means there are at least $i$ $C$ variables set false in the leaves that descend from $k$. Using this method requires adding $O(|G|)$ $b$ variables to the SAT implementation, which are not present in the ILP version, however—consistent with the ILP implementation— the number of variables required for the whole CNF formula remains $O(n|G|)$.

*Testing.* Both real and random sequences were tested for both 2-D and 3-D. The random sequences were generated with two different distributions: one where the average proportion of 1s in the sequence was $\frac{2}{3}$ (this matches the actual case of globular proteins); and one where the average proportion was $\frac{1}{3}$. The random sequences ranged in length from six to twenty-four characters long for 2-D and from six to sixteen characters long for 3-D. The same random sequences were used for both 2-D and 3-D and were only tested with Gurobi and pLingeling. Ten random sequences of each length from both distributions were generated for testing, so each length was tested with a total of twenty random sequences. The real sequences used are a subset of those described and used in [12] and ranged in length from six to twenty-nine for 2-D and from six to twenty-five for 3-D; all were tested using Gurobi, pLingeling, and Glucose-Syrup. All the sequences were tested five times in order to compare the variance in run times on the same sequence. The tests were done via a pipeline that generated the files for both the CNF and ILP files; passed the files to the SAT and ILP solvers; and recorded the number of contacts found and the time taken by each solver. All the tests were done using a 2018 Mac Mini with a 3.2 GHz i7 processor. The pipeline programs for both the 2-D and 3-D versions, along with the input sequences used can be found on GitHub.[2]

Unlike Gurobi, the SAT-solvers can only report whether a target number of potential contacts is possible or not. Because of this, it was necessary to include a wrapper binary-search function which performs a modified binary search over the targets. The search starts with target 1 and doubles the target until a result of *unsatisfiable* is returned, then executes a normal binary search between that value and the last satisfiable value. The times for each run of the SAT-solver are recorded and aggregated in the end report, but the *generation* of the CNF formulas was not timed, as it is minimal, and not the subject of this study.

*Results.* Overall, Gurobi performed better than both pLingeling and Glucose-Syrup, both achieving faster times to reach an optimal solution and having smaller files. Between the two SAT-solvers, pLingeling performed better than Glucose-Syrup. In the tests run on real sequences, Glucose-Syrup generally performed better on short sequences than pLingeling, but was outperformed by pLingeling for longer sequences in both the 2-D and 3-D versions. One counter-point of importance is that there were some sequences where Gurobi took much longer than did either SAT-solver,[3] and there were fewer cases where pLingeling took a similar amount more time than Gurobi.[4] Thus, pLingeling exhibited somewhat greater robustness than did Gurobi. Although both the SAT and ILP formulations have $O(n|G|)$ variables, it takes many more clauses and occurrences of un-negated and negated variables to describe the same problem using CNF than using ILP inequalities.

*Comparison of SAT and ILP Run Times.* The run times follow a trend that is similar to the trend for file sizes, with pLingeling and Glucose-Syrup generally taking longer than Gurobi for both 2-D and 3-D variants (see Fig. 1) and having a much higher variance per run than Gurobi. There was a small subset of problem instances that both SAT-solvers were able to solve much faster than Gurobi, but this was not the norm. For the 2-D version (with sequences between length 6 and length 29), pLingeling was able to reach a solution to most in less than 30 min, with none taking longer than three hours. Glucose-Syrup took at least twice as

---

[2] `prototein-problem` codebase: https://github.com/hannah-aught/prototein-prob lem.

[3] Notably, the `2drpD2`, `1tf3A1`, and `1a1iA1` sequences in the 2-D version.

[4] The `1be3k0` and `1byya0` sequences in the 3-D version, which, respectively, took pLingeling 3 h and half an hour to solve and Gurobi 5 min and 28 s.

long as pLingeling on most of the real sequences longer than 24, but performed slightly better on shorter sequences. Gurobi had times under ten minutes for most sequences between length 6 and 27 in 2-D; however, at lengths 28 and 29 the times became much less consistent, ranging from under 10 min to more than 48 h. Due to space limitations, we only show three graphs, two for real data in 2-D and 3-D, and one for random data in 2-D.

For the 3-D version (with lengths ranging from 6 to 15), all solvers were generally able to reach a solution within an hour, with Gurobi usually being the fastest. However, unlike in the 2-D version, Gurobi's times stayed more consistent at each length than both SAT-solvers'. At lengths 16 and above, all solvers became less consistent, with times between 20 s and 16 h for pLingeling, between 2 h and more than 3 days for Glucose-Syrup,[5] and between 3 s and 1.6 h for Gurobi.

*Conclusion from our Exploration of the HP Problem.* Gurobi generally performs better in terms of time, and the file sizes are always smaller. Although both SAT-solvers usually take longer than Gurobi on both 2-D and 3-D versions of the problem, there are (rarer) cases where Gurobi takes much longer than either on a particular sequence. Hence, if Gurobi seems to be taking much longer than the typical time for that length, it is worth trying pLingeling or Glucose-Syrup.[6]

## 3    Transforming Gene Order by Reversals

There are important biological phenomena that occur at a scale larger than individual nucleotides, e.g. at *chromosomal* or *genomic* scales. One such informative biological phenomenon is *long chromosomal reversals (inversions)*, where the DNA in a long interval on a chromosome *reverses* direction. For example, if we represent each gene by a distinct integer, the interval with ten genes: 1 10 4 5 2 6 3 9 8 7 becomes: 1 10 6 2 5 4 3 9 8 7 when the interval containing 4 5 2 6 is reversed. Since genes can be separated by long distances on a chromosome, what seems like a small reversal of just four integers actually represents a very long chromosomal reversal. It is believed that long chromosomal reversals are much rarer, with longer periods of time between reversals, than are mutations of single nucleotides, so long chromosomal reversals allow us to look *farther back* into evolutionary history. So, what we are interested in is the *order* of the genes on the chromosome, and how that order changes over time. Then, following the principle of *parsimony*, the computational problem of interest is:

**The Sorting-by-Reversals Problem.** Given two permutations, $P_1$ and $P_2$, of the integers 1 to $n$, find the *minimum* number of interval reversals that transforms $P_1$ to $P_2$.

---

[5] At which point it had still not reached a result and the process was terminated.

[6] Although pLingeling was seen to perform better than Glucose-Syrup for most sequences.

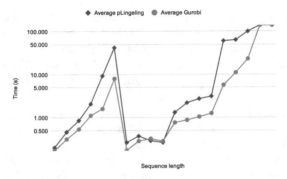

(a) Randomly generated sequences with $\frac{2}{3}$ ones in the 2-D version

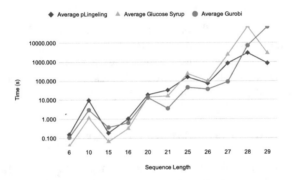

(b) Real sequences in the 2-D version

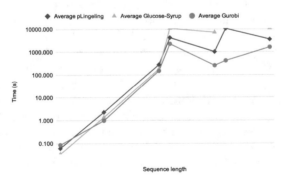

(c) Real sequences in the 3-D version

**Fig. 1.** Average SAT and ILP runtimes vs. sequence length for 2-D and 3-D embeddings. Note that the drop in times at lengths 12 and 20 for the 2-D and 3-D versions (respectively) is due to the choice of a smaller grid diameter

In the *signed* version of the problem, each integer in $P_1$ has a positive sign, and each integer in $P_2$ has either a positive or negative sign. Whenever an interval is reversed, the sign of each integer in the interval changes to the opposite sign. Then, the goal, as before, is to transform $P_1$ to the correctly signed $P_2$ using the minimum number of reversals.

*A Hard Problem.* There are no known algorithms to solve the sorting-by-reversals problem that are efficient in the worst-case sense,[7] and it is not obvious how to solve sorting-by-reversal problems in practice. It is even less obvious how one could efficiently *prove* that an optimal solution had been found, even if you had one. So, ILP formulations for this problem were previously created and examined in [10, 14, 23], and an exposition of the ILP approach in [14] appears in [8].

*ILP v. SAT.* Although the ILP approach successfully solves many datasets, we previously found that the ILP approach becomes impractical (on a 4-core MacBook Pro) on permutations longer than around twelve. Therefore, we investigated how well the SAT approach handles the problem of sorting-by-reversals, both the unsigned and signed versions.

Two classic datasets were first examined. In the **Turnips to Cabbage** dataset [22], with ten genes, the turnip genes were labeled in the natural order 1 through 10, so that the order of the analogous genes in Cabbage is 1 10 4 5 2 6 3 9 8 7.

Gurobi 8.1 solved the problem in fifteen seconds, and the SAT formulation was generated and solved by our program, `do-unsigned`, in five seconds. `do-unsigned.py` implements the SAT approach for a given instance of the reversal problem by creating a series of CNF formulas, each specifying a maximum allowed number of reversals. Each CNF formula is found to be satisfiable or unsatisfiable by the SAT-solver *pLingeling*.

Note that the ILP time only includes the solution time by the ILP solver, excluding the time (usually quite small) to create the ILP formulation from the description of a problem instance. In contrast, the time for the SAT formulation also includes the time to generate (by `do-unsigned.py`) the CNF formulas used during the search for the minimum number of reversals.

In the **Field Mustard to Black Mustard** dataset, of length twelve, the ILP formulation was solved by Gurobi 8.1 in about 102 min, but earlier in only 33 min by Gurobi 8.0 (go figure!). The SAT approach solved the reversal problem in about 3 min, 25 s.

These tests indicate overall that the SAT approach is superior to the ILP approach. Results from the examination of random permutations of length up to eight (shown in Table 1) is consistent with these tests.

We also compared running times for the ILP and SAT approaches to the *signed* variant. There, the overall running times were longer for both approaches, but more interestingly, there were substantial differences in the times for the two approaches, shown in Table 2, which show dramatic superiority of the SAT approach.

---

[7] Although there is an efficient algorithm for the *signed* version [9].

**Table 1.** Times (rounded to the nearest tenth of a second) taken by Gurobi and pLingeling with ten randomly generated unsigned sequences of length eight

| Sequence number | 1 | 2 | 3 | 4 | 5 | 6 | 7 | 8 | 9 | 10 |
|---|---|---|---|---|---|---|---|---|---|---|
| pLingeling | 1.3 | 3.1 | 2.5 | 1.3 | 1.5 | 1.7 | 2.8 | 1.6 | 12.9 | 1.3 |
| Gurobi | 5.0 | 13.5 | 11.2 | 9.0 | 6.2 | 10.5 | 9.1 | 3.0 | 74.0 | 8.7 |

**Table 2.** Times (rounded to the nearest second) taken by Gurobi and pLingeling with ten randomly generated signed sequences of length eight

| Sequence number | 1 | 2 | 3 | 4 | 5 | 6 | 7 | 8 | 9 | 10 |
|---|---|---|---|---|---|---|---|---|---|---|
| pLingeling | 57 | 463 | 56 | 11 | 50 | 430 | 54 | 58 | 19 | 78 |
| Gurobi | 30381 | 36387 | 7095 | 7855 | 3294 | 62050 | 6282 | 22293 | 362 | 13618 |

*Other Observations.* Although we saw that the SAT approach was faster than the ILP approach—and sometimes dramatically so—on permutations of length up to twelve, we found that its range of practicality is not much greater than that of the ILP approach. Another interesting finding is that for a fixed permutation length, the speed of the ILP approach falls as the number of cycles in the permutation increases, while the speed of the SAT approach is essentially constant over those permutations. For example, for length seven permutations, the SAT approach takes about one-half second no matter what the permutation is, but the time for the ILP approach falls from three seconds for permutations with one cycle, to about one-half second for permutations with four cycles.

*Software.* All the software for the SAT approach to the reversal problem is available on GitHub.[8]

## 4    The History Bound in Phylogenetic Networks

Genealogical and phylogenetic **networks** are graph-theoretic models of evolution that go beyond phylogenetic **trees**, incorporating *non-treelike* biological events such as *meiotic recombination* that occur in *populations* of individuals in a *single* species. The central algorithmic problems are to reconstruct *plausible* histories, with mutations, treelike events, and non-treelike events that generate a given set of observed genomic *sequences*, and to determine the *minimum* number of such biological events needed to derive the sequences.

The problem of finding a genealogical network that reconstructs the observed sequences, $M$, using the fewest recombinations (denoted $R_{min}(M)$), and at most one mutation per site, has been heavily studied in both population genetics and

---

[8] `tranform-order-by-reversals` codebase: https://github.com/zl674421351/transfo rm-order-by-reversals.

computational biology [7]. However, the best current algorithm takes super-exponential time in worst case. Hence, there has been great interest in finding effective algorithms to compute good (high) *lower bounds* on $R_{min}(M)$. Empirically, the *History Bound* is usually the highest lower bound of all the lower bounds studied (there are more than ten bounds that have been intensely studied). However, its computation, by dynamic programming [1], examines all $2^n$ subsets of $n$ input sequences, and hence becomes impractical as $n$ increases.[9] In its most intense use, in a program called *RecMin* [20], $n$ is very large, but the number of sites, $m$, is generally under ten.

*The Defining Algorithm for the History Bound.* The History Bound was first defined *procedurally*,[10] based on an algorithm that computes a *Candidate History Bound (CHB)*, given an $n$-by-$m$ binary matrix $M$. In short, the CHB algorithm reduces $M$ to the empty matrix by iterating three rules for removing rows or columns. In one rule (called the $D_t$ rule), the choice of row removal is *arbitrary*, so different executions of the algorithm, with the same input, can remove a different set of rows. The History Bound is defined as the minimum number of times rule $D_t$ is used, over *all* possible executions of Algorithm CHB. For a proof that the History Bound is a true lower bound on $R_{min}(M)$, see [7].

*The History Bound, ILP and SAT.* Previously, we developed three different ILP formulations to compute the History Bound [18]. None of these three ILP formulations was effective in empirical testing on data of current interest. Therefore, we developed a SAT approach for the History Bound, as follows.

For any given target, $t$, our program creates a CNF formula that is satisfiable if and only if there is an execution of Algorithm CHB on the input $M$, which applies rule $D_t$ at most $t$ times. The CNF formula encodes the three row and column removal rules of Algorithm *CHB*, with additional CNF clauses that "count" the number of times rule $D_t$ is used; and has another clause to forbid using it more than $t$ times. Then pLingeling determines if that CNF formula is satisfiable or not. The History Bound is found by searching over the range of possible target values to find the smallest target where the associated CNF is satisfiable.

*Results.* When the target $t$ is larger than the History Bound,[11] so that the associated CNF formula is satisfiable, pLingeling very efficiently finds a satisfying assignment. However, pLingeling has difficulty terminating when the CNF formula is *not* satisfiable. Space limits us to a few illustrative examples. For a matrix of dimension 60-by-8, the SAT approach determined that the History Bound is less than 11 in 45 s; less than 10 in 64 s; less than 9 in 31 s; less than 8 in 52 s; and less than 7 in 87 s. Then the test of whether the History Bound is less than

---

[9] The original algorithm, in [20] required $\theta(mn!)$ time, where each of the $n$ input sequences has $m$ sites.

[10] A non-procedural definition of the History Bound was later established in [19].

[11] In tests where we know the true History Bound.

6 ran for 7,200 s without termination, at which point we killed the computation. In contrast, the ILP approach determined that the History Bound is less than 11 after 900 s, and less than 7 after 3,351 s; and that computation continued without termination until 18,000 s, when it was killed. Significantly, the ILP could only establish that the History Bound for this case is greater or equal to zero! So, the SAT approach found an execution of Algorithm CHB using the $D_t$ rule only six times, much faster than did the ILP approach. And it provided much stronger evidence than did the ILP computation, for concluding that the History Bound is six for that matrix.

In a test with a matrix of dimension 30-by-60, the SAT approach determined that the History Bound is less than 30, 29, 28, and 27, in 21, 38, 26, and 63 s respectively. Then in trying to determine if the History Bound is less than 26, we terminated pLingeling after 480 s. In contrast, we allowed the ILP computation to run for 13,000 s at which point it has determined only that the History Bound was less than 30 and greater than zero! So, the SAT approach gave strong evidence that the History Bound is 26 for this matrix, while the ILP approach gave much less information, using much more time.

All of the software to create the ILP and CNF formulations for the History Bound is available on GitHub.[12]

# 5   Haplotyping by Pure Parsimony

Very briefly, input consists of $n$ sequences of length $m$, over a *ternary* alphabet $\{0, 1, 2\}$. Output consists of a set of binary sequences, $H$, of length $m$, where each input sequence, $s$, is associated with some pair of binary sequences, $(h_i, h_j)$ in $H$, which must satisfy the following conditions: At any position $k$ in $s$, if the value of $s$ is 0 or 1, then for both $h_i$ and $h_j$, the value at position $k$ must be identical to the value in $s$. However, if the value of $s$ at position $k$ is 2, then exactly one of $h_i$ or $h_j$ must have value 0, and the other must have value 1, at position $k$. When $s$ has $q$ positions with value 2, there are $2^{q-1}$ possible binary sequences that satisfy those conditions.

The **Pure-Parsimony Haplotyping Problem** is to find a *smallest* such set $H$, given the input. This minimization problem was first examined in [6], where an exponential-size ILP formulation was developed.[13] That ILP formulation solves very quickly for instances where the formulation fits into main memory. However, rapid growth of the formulation size limits the size of the instances that can be handled. Later (in 2004 and 2006), polynomial-sized ILP formulations were developed [4,13], but ILP solvers at that time were not able to solve problem instances of interest in a practical amount of time. Subsequently (in 2006), a SAT-formulation, based on the general ideas in [4,13] and additional improvements, was developed in [15,16], and was shown to be highly

---

[12] `History Bound/Haplotyping` codebase: https://github.com/gusfield/ILP-SAT-com parisons.

[13] See also [8] for a full discussion of the problem and its biological import.

effective. Those results used ILP and SAT-solvers that are much slower than current solvers. So, we re-implemented the ILP and SAT approaches[14] to see how the more modern SAT-solver, pLingeling, and the fastest current ILP-solver, Gurobi 9.0, handle this problem. The code for our implementation is available on GitHub.[15]

Our new results confirm and extend the results from the older studies. Space limits the discussion to just one problem instance, of dimension 200-by-200, which is much larger than the instances examined in the earlier literature. In the SAT approach, using a first target of 87, pLingeling found a satisfying solution to the CNF formula in 1,525 s. With targets of 85 and 84, it found satisfying solutions in 5,700 and 8,521 s, respectively. Then, using a target of 83, pLingeling determined that the associated CNF formula was unsatisfiable, in 20,627 s. This established that the optimal value is 84. In contrast, using the ILP formulation for this problem instance, Gurobi initially found a feasible solution of size 134, and a lower bound of 31 (in 1,300 s), but then made no further progress in two more days of computation (when it was terminated).

## 6 Conclusions

While the ILP approach is becoming more widely used in computational biology, only a few studies have explored the SAT-solving approach. Our work shows that SAT-solving can be (but is not always) much more effective than ILP-solving for instances of hard problems in computational biology.[16] This gives us another powerful tool, which should be more widely used and incorporated into biological computation.

## References

1. Bafna, V., Bansal, V.: Inference about recombination from haplotype data: lower bounds and recombination hotspots. J. Comput. Biol. **13**, 501–521 (2006)
2. Bailleux, O., Boufkhad, Y.: Efficient CNF encoding of Boolean cardinality constraints. In: Rossi, F. (ed.) CP 2003. LNCS, vol. 2833, pp. 108–122. Springer, Heidelberg (2003). https://doi.org/10.1007/978-3-540-45193-8_8
3. Biere, A.: http://fmv.jku.at/lingeling
4. Brown, D., Harrower, I.: Integer programming approaches to haplotype inference by pure parsimony. IEEE/ACM Trans. Comput. Biol. Bioinform. **3**(2), 141–154 (2006)

---

[14] Although our SAT implementation is more basic and naive than the one developed in [15,16].

[15] `History Bound/Haplotyping` codebase: https://github.com/gusfield/ILP-SAT-comparisons.

[16] We reiterate, however, that the specific problems examined in this paper were chosen because they were known to be hard for ILP-solvers to handle. Therefore, these results do not contradict the more general conclusion (such as detailed in [8]), that the ILP approach has broad, successful (even transformative) application in computational biology.

5. Dill, K.A., et al.: Principles of protein folding - a perspective from simple exact models. Protein Sci. **4**, 561–602 (1995)
6. Gusfield, D.: Haplotype inference by pure parsimony. In: Baeza-Yates, R., Chávez, E., Crochemore, M. (eds.) CPM 2003. LNCS, vol. 2676, pp. 144–155. Springer, Heidelberg (2003). https://doi.org/10.1007/3-540-44888-8_11
7. Gusfield, D.: ReCombinatorics: The Algorithmics of Ancestral Recombination Graphs and Explicit Phylogenetic Networks. MIT Press, Cambridge (2014)
8. Gusfield, D.: Integer Linear Programming in Computational and Systems Biology: An Entry-Level Text. Cambridge University Press, Cambridge (2019)
9. Hannenhalli, S., Pevzner, P.: Transforming cabbage into turnip: polynomial algorithm for sorting signed permutations by reversals. In: Proceedings of the 27th ACM Symposium on the Theory of Computing, pp. 178–189 (1995)
10. Hartmann, T., Wieseke, N., Sharan, R., Middendorf, M., Bernt, M.: Genome rearrangement with ILP. IEEE/ACM Trans. Comput. Biol. Bioinform. **15**(5), 1585–1593 (2018). https://doi.org/10.1109/TCBB.2017.2708121
11. Knuth, D.E.: The Art of Computer Programming. Fascicle 6: Satisfiability, vol. 4. Addison-Wesley, Boston (2015)
12. Kolodny, R., Koehl, P., Levitt, M.: Comprehensive evaluation of protein structure alignment methods: scoring by geometric measures. J. Mol. Biol. **346**(4), 1173–1188 (2005). https://doi.org/10.1016/j.jmb.2004.12.032
13. Lancia, G., Pinotti, C., Rizzi, R.: Haplotyping populations by pure parsimony: complexity, exact and approximation algorithms. INFORMS J. Comput. Spec. Issue Comput. Biol. **16**, 348–359 (2004)
14. Lancia, G., Rinaldi, F., Serafini, P.: A unified integer programming model for genome rearrangement problems. In: Ortuño, F., Rojas, I. (eds.) IWBBIO 2015. LNCS, vol. 9043, pp. 491–502. Springer, Cham (2015). https://doi.org/10.1007/978-3-319-16483-0_48
15. Lynce, I., Marques-Silva, J.: Efficient haplotype inference with Boolean satisfiability. In: Proceedings of the Twenty-First AAAI Conference on Artificial Intelligence, pp. 104–109 (2006)
16. Lynce, I., Marques-Silva, J.: SAT in bioinformatics: making the case with haplotype inference. In: Biere, A., Gomes, C.P. (eds.) SAT 2006. LNCS, vol. 4121, pp. 136–141. Springer, Heidelberg (2006). https://doi.org/10.1007/11814948_16
17. Malikic, S., et al.: PhISCS: a combinatorial approach for subperfect tumor phylogeny reconstruction via integrative use of single-cell and bulk sequencing data. Genome Res. **29**, 1860–1877 (2019)
18. Matsieva, J.: Optimization techniques for phylogenetics. Ph.D. thesis, Department of Computer Science, University of California, Davis (2019)
19. Matsieva, J., Kelk, S., Scornavacca, C., Whidden, C., Gusfield, D.: A resolution of the static formulation question for the problem of computing the history bound. IEEE/ACM Trans. Comput. Biol. Bioinf. **14**(2), 404–417 (2017). https://doi.org/10.1109/TCBB.2016.2527645
20. Myers, S., Griffiths, R.C.: Bounds on the minimum number of recombination events in a sample history. Genetics **163**, 375–394 (2003)
21. Nunes, L., Galvao, L., Lopes, H., Moscato, P., Berretta, R.: An integer programming model for protein structure prediction using the 3D-HP side chain model. Discret. Appl. Math. **198**, 206–214 (2016)
22. Palmer, J., Herbon, L.: Plant mitochondrial DNA evolves rapidly in structure, but slowly in sequence. J. Mol. Evol. **27**, 87–97 (1988)
23. Shao, M., Moret, B.M.E.: Comparing genomes with rearrangements and segmental duplications. Bioinformatics **31**(12), i329–i338 (2015)

# Combining Networks Using Cherry Picking Sequences

Remie Janssen[1]([⊠]) [ID], Mark Jones[2] [ID], and Yukihiro Murakami[1] [ID]

[1] Delft Institute of Applied Mathematics, Delft University of Technology,
Van Mourik Broekmanweg 6, 2628 XE Delft, The Netherlands
{R.Janssen-2,Y.Murakami}@tudelft.nl
[2] Centrum Wiskunde & Informatica,
P.O. Box 94079, 1090 GB Amsterdam, The Netherlands
markelliotlloyd@gmail.com

**Abstract.** Phylogenetic networks are important for the study of evolution. The number of methods to find such networks is increasing, but most such methods can only reconstruct small networks. To find bigger networks, one can attempt to combine small networks. In this paper, we study the NETWORK HYBRIDIZATION problem, a problem of combining networks into another network with low complexity. We characterize this complexity via a restricted problem, TREE-CHILD NETWORK HYBRIDIZATION, and we present an FPT algorithm to efficiently solve this restricted problem.

**Keywords:** Phylogenetic networks · Network hybridization ·
Tree-child networks · FPT algorithms

## 1 Introduction

Evolutionary histories are often represented by phylogenetic trees, and more recently, by phylogenetic networks. Knowing the evolutionary history of a species is vital for understanding their biology. Therefore, it is important to have methods for finding phylogenetic networks that accurately represent the true evolutionary histories. Many methods exist to find evolutionary histories; some are purely combinatorial, others have a likelihood component as well. Here, we focus mainly on the purely combinatorial problems.

One classic combinatorial method is to solve HYBRIDIZATION: given a set of trees, find the simplest network that displays these trees [1]. Unfortunately, the problem is NP-hard, even on inputs of two trees [2]. For this problem, it is assumed we can construct accurate phylogenetic trees for small parts of the genomes of the studied taxa. When the input consists of only two or three trees, it can be solved relatively efficiently—EPT time [8,17]—even though the problem

Research funded by the Netherlands Organization for Scientific Research (NWO), with the Vidi grant 639.072.602.

is already NP-hard in that case. For an input consisting of three trees or more, there is still an FPT algorithm [9], but it is not practical. In these cases, it is useful to limit the search space to networks with a restricted structure, such as tree-child networks [7], or temporal networks [6].

Another combinatorial approach for finding phylogenetic networks is to combine smaller networks. The smaller networks are often assumed to have certain properties. For example, it may be assumed that we are given all strict subnetworks containing the full set of leaves. In that case, it is possible to reconstruct a level-$k$ tree-child network from all its level-$(k-1)$ subnetworks [15]. Another assumption could be that the input consists of all subnetworks obtained by removing exactly one leaf [11]. Instead of having almost all leaves, the subnetworks can also be allowed to have only few leaves. For example, low level networks can be reconstructed from their full set of binets [5,12], trinets [10,16] or quarnets [4].

In practice, it may not be easy to find *all* subnetworks. This renders many of the previously mentioned methods useless. Furthermore, these methods typically only work for low level networks. This means that they cannot be used when the phylogenetic signal comes from a complicated evolutionary history, or if there is some randomness in the data, complicating the data as well.

In this paper, we combine networks that all contain the full set of leaves, but we do not assume we have all the subnetworks of the network we want to find. The problem we solve is analogous to HYBRIDIZATION, but with networks as the input, NETWORK HYBRIDIZATION: Given a set of networks with taxa $X$, find a network $N$ with minimal reticulation number, that displays all input networks. Since this is a generalization of the HYBRIDIZATION problem, the problem remains NP-hard in general, even for inputs of two networks. We show that for the restricted problem on tree-child (topologically restricted class of networks; defined formally in Sect. 2) inputs and output, we can use tree-child sequences to obtain an FPT algorithm. This FPT algorithm is an extension of the one introduced in [7] in which they considered tree inputs; the tree-child sequence approach was first introduced in [14]. We also comment briefly on how some measure of an optimal solution to the NETWORK HYBRIDIZATION problem can be characterized by solving this restricted problem.

*Structure of the Paper.* We start with a quick introduction of relevant concepts from mathematical phylogenetics in Sect. 2. Then, in Sect. 3, we formally introduce TREE-CHILD NETWORK HYBRIDIZATION and prove its relation to tree-child sequences. This section also relates this problem to the more general NETWORK HYBRIDIZATION. In Sect. 4.1, we lay the theoretical foundation to extend the algorithm in [7] that takes inputs of trees to also work for inputs of networks. As a last theoretical section in the paper, we present an FPT algorithm that solves TREE-CHILD NETWORK HYBRIDIZATION (Sect. 4.2). We conclude the paper with a discussion, including some open questions (Sect. 5).

## 2    Preliminaries

The main objects of study for this paper are phylogenetic networks. These graphs are used in biology to represent evolutionary scenarios for a given set of species.

**Definition 1.** *A* (rooted phylogenetic) network *on a finite set of* taxa $X$ *is a directed acyclic graph with*

- *one node of indegree-0 and outdegree-1, the* root;
- *nodes of indegree-1 and outdegree-0 labelled bijectively with* $X$, *the* leaves;
- *nodes of indegree-1 and outdegree-2, the* tree nodes;
- *nodes of indegree greater than 1 and outdegree-1, the* reticulations.

If all the reticulation nodes have indegree-2, the network is called *binary*. An edge $uv$ is called a *tree edge* if $v$ is a tree node or leaf, and a *reticulation edge* if $v$ is a reticulation. The vertex $u$ is the *parent* of $v$, and $v$ is the *child* of $u$. The *reticulation number* $r(N)$ of a network $N$ is the total number of reticulation edges minus the total number of reticulations.

A network is *stack-free* if every reticulation has a child that is a tree node or a leaf. A network is *tree-child* if it is stack-free and every tree node has a child that is a tree node or a leaf. We now define some relevant notation for local structures in phylogenetic networks.

**Definition 2.** *Let $N$ be a network on $X$ and $x, y \in X$ two leaves. Then we say $N$ has a* cherry $\{x, y\}$ *if the parent of $x$ is the parent of $y$; we say that $N$ has a* reticulated cherry $(x, y)$ *if the parent of $x$ is a reticulation, and the parent of $y$ is a parent of this reticulation. If $(x, y)$ is a cherry or a reticulated cherry in $N$, then it is called a* reducible pair.

Tree-child sequences are built on the notion of *reducing* cherries and reticulated cherries from networks.

**Definition 3.** *Let $N$ be a network on $X$, and $(x, y)$ a pair of leaves. Let $p_x$ and $p_y$ denote the parents of $x$ and $y$ in $N$, respectively Then* reducing $(x, y)$ *in $N$ results in a network $N(x, y)$ obtained as follows:*

- *If $\{x, y\}$ is a cherry in $N$, remove $x$ and the pendant edge $p_x x$, and suppress $p_x$ if it has become a degree-2 node;*
- *If $(x, y)$ is a reticulated cherry in $N$, remove the reticulation edge $p_y p_x$ and suppress $p_x$ or $p_y$ if it has become a degree-2 node.*
- *Otherwise, $N(x, y) := N$.*

The reversal of reducing cherries and reticulated cherries can be done by *adding* ordered pairs of leaves to the network.

**Definition 4.** *Let $N$ be a network and let $(x, y)$ be reducible pair. Then we may construct $N$ from $N(x, y)$ —also called* add $(x, y)$ to $N(x, y)$ —*by applying the following.*

1. *If $x$ is a leaf in $N(x, y)$ (i.e., if $(x, y)$ is a reticulated cherry in $N$), and*
   (a) *if $p$, the parent of $x$ in $N(x, y)$, is a reticulation then add a node $q$ directly above $y$, and add an edge $qp$.*
   (b) *otherwise, add nodes $p$ and $q$ directly above $x$ and $y$ respectively, and add an edge $qp$.*
2. *If $x$ is not a leaf in $N(x, y)$ (i.e., if $(x, y)$ is a cherry in $N$) then add a labelled node $x$, insert a node $q$ directly above $y$, and add an edge $qx$.*

The above notion of adding an ordered pair of leaves $(x, y)$ to a network $N$ is well-defined if $y$ is already a leaf in $N$. If this is indeed the case, we may obtain a network from a sequence of ordered pairs by iteratively adding ordered pairs to an existing network. To do so, we impose the following condition on our sequence of ordered pairs: *The second coordinate of each pair has to occur as a first coordinate in the remainder of the sequence, or as the second coordinate of the last pair.* Then, the following procedure constructs a network from a sequence.

---

**Procedure** ConstructNetworkFromSequence($S$)

---

**Input**: A sequence of ordered pairs $S = (x_1, y_1) \cdots (x_n, y_n)$;
**Output**: The network that can be constructed from $S$;

1  Set $N$ to be the tree on one leaf $y_n$;
2  **for** $i = n, \ldots, 1$ **do**
3      **if** $x_i$ *is a leaf of $N$* **then**
4          **if** *the parent of $x_i$ is a reticulation* **then**
5              Let $p_x$ denote the parent of $x_i$;
6          **else**
7              Subdivide the incoming edge of $x_i$ with a node $p_x$;
8          Subdivide the incoming edge of $y_i$ with a node $p_y$;
9          Add the edge $p_y p_x$ to $N$;
10     **else**
11         Subdivide the incoming edge of $y_i$ with a node $p_y$;
12         Add a new node labelled $x_i$ to $N$;
13         Add the edge $p_y x_i$ to $N$;
14 **return** $N$;

---

Note that because we only add reticulation edges to existing reticulation nodes wherever possible (Line 4), the network obtained by using the above procedure is always stack-free. Imposing another condition: *no first coordinate leaf is used as a second coordinate in the remainder of the sequence* on the sequence ensures that the network we obtain is tree-child. With this in mind, we formally define a tree-child sequence (Fig. 1).

**Definition 5.** *A tree-child sequence (TCS) is a sequence of ordered pairs of two leaves such that the following conditions hold:*

- *the second coordinate of each pair has to occur as a first coordinate in the remainder of the sequence or as the second coordinate of the last pair;*
- *no first coordinate leaf is used as a second coordinate in the rest of the sequence.*

Let $S$ be a TCS, that involves the leaves $X$. Then, the *weight* of $S$ is $w(S) = |S| - |X| + 1$. Given a sequence of ordered pairs $S = S_1 S_2 \cdots S_{|S|}$, we let $NS$ denote the network

$$NS := (\cdots ((NS_1)S_2) \cdots S_{|S|-1})S_{|S|} = NS_1 S_2 \cdots S_{|S|}.$$

We introduce some notation for subsequences of a sequence $S$. For $i \in [|S|]$, we use the following notation for subsequence of $S$. The $i$th ordered pair of $S$ is $S_i = (x_i, y_i)$. The first $i$ ordered pairs in $S$ is denoted by $S_{[:i]} = (x_1, y_1), \ldots, (x_i, y_i)$. The subsequence of $S$ without the first $i$ ordered pairs is denoted by $S_{[i+1:]} = (x_{i+1}, y_{i+1}), (x_{i+2}, y_{i+2}), \ldots, (x_n, y_n)$. We say that the leaves $x_1, \ldots, x_i$ are *forbidden* for $S_{[:i]}$. Forbidden leaves do not appear as a second coordinate leaf in a TCS (by the second condition of TCSs).

We say $S$ *reduces $N$ to the leaf $x$* if $NS$ is the tree with the single leaf $x$. Similarly, let $\mathcal{N}$ be a set of networks, then we denote by $\mathcal{N}S$ the set of reduced networks $\{NS : N \in \mathcal{N}\}$, and we say $S$ *reduces $\mathcal{N}$ to $x$* if $NS$ is the one leaf tree $x$ for all $N \in \mathcal{N}$.

We call a sequence $S'$ of ordered pairs a *partial TCS* if there exists a TCS $S$ such that $S_{[:i]} = S'$ for some $i$.

## 3    NETWORK HYBRIDIZATION

In this section we formally define the TREE-CHILD NETWORK HYBRIDIZATION problem, which asks to find a tree-child network with minimal reticulation

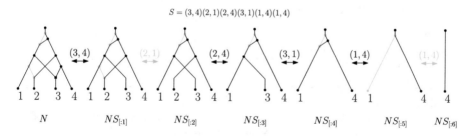

**Fig. 1.** A binary tree-child network $N$ (grey and black) reduced to a leaf 4 by a tree-child sequence $S$. The reduction is shown as a sequence of networks $NS_{[:i]}$ for $i = 0, 1, \ldots, 6$ from left to right, in which an element of $S$ is applied to the network successively. Each element of $S$ reduces a pair in $N$. An example of a cherry $(3, 1)$ can be seen in the network $NS_{[:3]}$, and a reticulated cherry $(3, 4)$ can be seen in the network $N$. The subnetwork of $N$ consisting of the black edges is also reduced by $S$, and the embedding can be constructed by building both networks simultaneously and keeping track of the edges added by the pairs that change the subnetwork (black pairs and arrows).

number that *displays* all input tree-child networks on the same set of taxa. We generalize the results presented in [14] (they considered inputs of trees while we consider inputs of networks) by showing how this problem relates to the more generalized problem of NETWORK HYBRIDIZATION and also to the TREE-CHILD WEIGHT problem. For the TREE-CHILD NETWORK HYBRIDIZATION problem, it turns out that there is not always a solution for some given inputs; we also comment on when this is the case.

We start by defining what it means for a network to *display* another network.

**Definition 6.** *Let $N$ be a network on the set of taxa, $X$. A network $N'$ on $Y \subseteq X$ is a* subnetwork *of $N$ if $N'$ can be obtained from $N$ by deleting reticulation edges, removing leaves not labelled by $Y$, and suppressing all degree-2 nodes in the resulting subgraph. If $N'$ can be obtained from a subnetwork of $N$ by contracting edges, then we say $N$* displays *$N'$. Given a set of networks $\mathcal{N}$ on some subsets of the taxa $X$, then we say that $N$* displays *$\mathcal{N}$ if $N$ displays all networks in $\mathcal{N}$.*

If a network $N'$ on $X$ is a subnetwork of another network $N$ on $X$, then an *embedding* of $N'$ into $N$ is the mapping of the nodes of $N'$ to the nodes of $N$ such that the leaves of $N'$ are mapped to the leaves of $N$ with the same labels, and the edges of $N'$ are mapped to edge-disjoint paths of $N$. Our main focus of this paper is to solve the following problem.

---

TREE-CHILD NETWORK HYBRIDIZATION
**Input:** A set of rooted tree-child networks $\mathcal{N}$ on $X$.
**Output:** A tree-child network that displays $\mathcal{N}$ with minimal reticulation number if it exists, NO otherwise.

---

Given an optimal tree-child network to the TREE-CHILD NETWORK HYBRIDIZATION problem, one may find a TCS that reduces it. We will show that the weight of such a TCS is equal to the weight of an optimal solution to the following related problem.

---

TREE-CHILD WEIGHT
**Input:** A set of rooted networks $\mathcal{N}$ on $X$.
**Output:** A minimal weight TCS that reduces $\mathcal{N}$ if it exists, NO otherwise.

---

Let $\mathcal{N}$ be a set of networks on $X$. The reticulation number of an optimal solution to TREE-CHILD HYBRIDIZATION is denoted $h_{\text{tc}}(\mathcal{N})$. The weight of an optimal solution to TREE-CHILD WEIGHT is denoted $s_{\text{tc}}(\mathcal{N})$.

For a set of trees $\mathcal{T}$, the relation $h_{\text{tc}}(\mathcal{T}) = s_{\text{tc}}(\mathcal{T})$ holds. We will extend this result for network inputs. We first recall some key lemmas from [13]. The first lemma loosely states that each TCS reducing a set of networks $\mathcal{N}$ gives a tree-child network with corresponding reticulation number that displays $\mathcal{N}$. The second lemma gives the reverse statement: each tree-child network that displays a set of networks $\mathcal{N}$ gives a TCS of corresponding weight that reduces $\mathcal{N}$.

**Lemma 7** ([13], **Lemma 8**). *Let $N$ and $N'$ be a tree-child network. Suppose there is a TCS $S$ that reduces both $N$ and $N'$, such that each element of $S$ that* *reduces a pair in $N'$ also reduces a reducible pair in $N$. Then $N'$ is a displayed by $N$ (Fig. 1).*

**Lemma 8** ([13], **Corollary 4**). *Let $N, N'$ be tree-child networks on $X$ such that $N'$ is displayed by $N$. If a TCS $S$ reduces $N$, then $S$ also reduces $N'$.*

Unlike when the input consists of only trees, a solution to TREE-CHILD NETWORK HYBRIDIZATION does not always exist when the input may also contain networks (Fig. 1).

**Definition 9.** *A set of tree-child networks $\mathcal{N}$ are* tree-child compatible *if there exists a tree-child network that displays all networks in $\mathcal{N}$.*

Our next step, is to prove that there is a strong connection between tree-child compatibility and TCSs.

**Lemma 10.** *Let $\mathcal{N}$ be a set of tree-child networks on $X$. Then $\mathcal{N}$ is tree-child compatible iff there exists a TCS that reduces $\mathcal{N}$. Furthermore, if a solution exists, then $h_{\mathrm{tc}}(\mathcal{N}) = s_{\mathrm{tc}}(\mathcal{N})$.*

*Proof.* Suppose that $\mathcal{N}$ is tree-child compatible. Then there exists a tree-child network $N$ that displays $\mathcal{N}$, with minimal reticulation number. Now let $S$ be a TCS for $N$. By Lemma 8, $S$ also reduces all displayed networks of $N$, and hence it reduces $\mathcal{N}$. Furthermore, the weight of $S$ is equal to the reticulation number of $N$ by Lemma 3 from [13], (originally proved slightly less strong in [14]).

Now suppose there exists a TCS $S$ that reduces $\mathcal{N}$. Let $N$ be the tree-child network that can be constructed from $S$. Then, by Lemma 7, $N$ displays $\mathcal{N}$. Because $N$ is the network corresponding to $S$, the reticulation number of $N$ is equal to the weight of $S$. □

### 3.1   The Existence of a Tree-Child Solution

In the previous subsection, we have found a strong connection between TREE-CHILD NETWORK HYBRIDIZATION and TREE-CHILD WEIGHT for feasible solutions. Not all inputs, however, are feasible. Here, we investigate the feasibility of inputs, and how to deal with infeasible inputs.

**Lemma 11.** *Let $N$ be a tree-child network with reticulated cherry $(x, y)$, then any TCS that reduces $N$ must contain the pair $(x, y)$.*

*Proof.* Suppose $S$ is a TCS that reduces $N$. The only ways to reduce the reticulated cherry $(x, y)$ are by either reducing it directly with the pair $(x, y)$, or by first turning it into a cherry $\{x, y\}$ and then reducing it with a pair $(x, y)$ or $(y, x)$. This second option, however, leads to a contradiction: To make the reticulated cherry into a cherry, we must reduce a pair of the form $(x, \cdot)$; however, any sequence that includes $(x, \cdot)$ and later $(y, x)$ cannot be tree-child. □

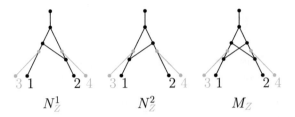

**Fig. 2.** Two networks $\mathcal{N} = \{N^1, N^2\}$ that are tree-child incompatible (black parts only). The network $M$ displays $\mathcal{N}$, but it is not tree-child. By adding leaves $Z = \{3, 4\}$ to $M$, we get the network $M_Z$ which is tree-child. Then, adding these leaves in the right places to $N^1$ and $N^2$, we get the set of networks $\mathcal{N}_Z \in \mathcal{N}^+$ on $X \cup Z$, that are displayed by the tree-child network $M_Z$.

Using the connection between tree-child compatibility and the existence of TCSs, we can prove an obstruction to tree-child compatibility of Lemma 12. This obstruction will turn out to be quite valuable in the proofs in the rest of this paper, as it allows us to quickly check whether a set of networks is tree-child compatible.

**Lemma 12.** *Let $N_1, N_2$ be tree-child networks on the same set of leaves $X$. For any pair of leaves $x, y$, if $N_1$ contains the reticulated cherry $(x, y)$ and $N_2$ contains the reticulated cherry $(y, x)$, then $N_1$ and $N_2$ are not tree-child compatible.*

*Proof.* Let $N$ be a tree-child network that displays both $N_1$ and $N_2$. Then any TCS $S$ for $N$ reduces both $N_1$ and $N_2$. By Lemma 11, the sequence $S$ must contain the pair $(x, y)$, because $N_1$ has the reticulated cherry $(x, y)$; similarly, $S$ must contain $(y, x)$. This means $S$ is a TCS, but it includes both pairs $(x, y)$ and $(y, x)$, a contradiction. Hence we conclude that $N_1$ and $N_2$ are not tree-child compatible. □

Even if an input is infeasible, we still desire a network that displays all input networks. For this purpose, we can relax the tree-child constraint on output (and input) of the TREE-CHILD NETWORK HYBRIDIZATION problem, giving rise to the following problem.

---
NETWORK HYBRIDIZATION
**Input:** A set of rooted networks $\mathcal{N}$ on $X$.
**Output:** A network that displays $\mathcal{N}$ with minimal reticulation number.

---

This problem can be viewed as the natural extension of the classic HYBRIDIZATION problem for trees. Linz and Semple show that HYBRIDIZATION can be solved by adding leaves in the right place to all input trees, and then solving TREE-CHILD HYBRIDIZATION [14]. This also holds for the network versions of these problems, as the solution to NETWORK HYBRIDIZATION can be made tree-child by adding leaves, and all networks displayed by a tree-child network are tree-child networks as well (Fig. 2).

# 4   An Algorithm for TREE-CHILD NETWORK HYBRIDIZATION

In this section, we give an FPT algorithm for TREE-CHILD NETWORK COMBINATION. We extend the algorithm given in [7] by allowing for inputs to be networks, and by looking for reducible pairs within networks rather than cherries in trees. Given an input $\mathcal{N}$ of tree-child networks, we first look for *trivial* reducible pairs. We show that it is safe to reduce trivial reducible pairs as soon as we encounter one, in any order. We then branch on all possible non-trivial reducible pairs of the network, and by showing that the total number of possible reducible pairs at each branching point is at most $8k$ for the reticulation number $k$ of the optimal solution, we show that the running time of the algorithm is $O((8k)^k \cdot \mathrm{poly}(|X|, |\mathcal{N}|))$.

## 4.1   Counting Cherries

**Trivial Pairs.** The algorithm in [7] reduces *trivial cherries* (a pair of leaves $\{x, y\}$ that appear as a cherry in any input tree containing $x$ and $y$) whenever possible. Here, only looking at trivial cherries is not sufficient. For an input of networks, we will need to reduce *trivial reducible pairs* (trivial pairs for short) whenever possible. A trivial pair is a pair of leaves $(x, y)$ such that all networks either only have the leaf $y$, or they have a reducible pair $(x, y)$. In the following two lemmas, we prove that it is safe to reduce such a pair as soon as we encounter one.

**Lemma 13 (Move to the left).** *Let $\mathcal{N} = \{N_1, \dots, N_I\}$ be a set of tree-child networks on a common set of leaves, and let $S(a, b)(x, y)S'$ be a TCS for $\mathcal{N}$. Suppose that for each $N \in \mathcal{N}S$ we have either $x$ is not a leaf in $N$, or $(x, y)$ is a reducible pair of $N$, and there is at least one network such that the latter holds. Then there is a TCS for $\mathcal{N}$ starting with $S(x, y)$ of length equal to that of $S(a, b)(x, y)S'$.*

*Proof.* Suppose $b = x$. Then there must be a network in $\mathcal{N}S$ that has both the reducible pairs $(x, y)$ and $(a, x)$. This can only occur if $a = y$: as $x$ is the first as well as the second element of a reducible pair, it must form a cherry with another leaf, namely the leaf $y$. However, $S(y, x)(x, y)S'$ is not a TCS, which contradicts our assumption that $S(a, b)(x, y)S'$ is a TCS for $\mathcal{N}$.

Hence, for the rest of the proof, we assume $b \neq x$. In this case, $S(x, y)(a, b)S'$ is a TCS. It remains to prove that it reduces $\mathcal{N}$. This is clear if $\{x, y\} \cap \{a, b\} = \emptyset$. Observe that $a \neq y$, as otherwise $S(a, b)(x, y)S'$ would not have been a TCS to begin with. Therefore, we still need to check the cases $a = x$ and $b = y$.

If $a = x$ and a network has both reducible pairs $(x, b)$ and $(x, y)$, then this network has a reticulation with reticulated cherries $(x, b)$ and $(x, y)$. The order of reducing these pairs obviously does not matter for such networks: both options remove the reticulation edges between the parents of $b$ and $y$, and the parent of $x$. For a network $N$ that only has the reducible pair $(x, y)$ after $S$ (and not

$(x, b))$, the network $NS(x, y)(x, b)$ is a subnetwork of $NS(x, b)(x, y) = NS(x, y)$. This means $S(x, y)(x, b)S'$ also reduces $N$ [13]. Hence if $a = x$, the sequence $S(x, y)(a, b)S'$ is a TCS for $\mathcal{N}$.

Now suppose $b = y$. Let $N$ be a network that has both reducible pairs $(a, y)$ and $(x, y)$. But all tree nodes of $N$ are of outdegree-2; this implies that every leaf can be the second coordinate of at most one reducible pair. Therefore such a network cannot exist, and thus this case is not possible.  □

**Lemma 14 (Trivial pair reduction).** *Let $\mathcal{N} = \{N_1, \ldots, N_I\}$ be a set of tree-child networks on a common set of leaves such that there exists a TCS $SS'$ for $\mathcal{N}$. Suppose $x, y$ are leaves such that for each $N \in \mathcal{N}S$ we have either $x$ not in $N$, or $(x, y)$ a reducible pair of $N$, and there is at least one network such that the latter holds. Then there exists a TCS $S(x, y)S''$ of length equal to $SS'$ that reduces $\mathcal{N}$, or if $y$ is forbidden after $S$ and there is a sequence of the form $S(y, x)S'''$ of the same length as $SS'$ that reduces $\mathcal{N}$.*

*Proof.* To reduce a network with reducible pair $(x, y)$, the sequence $S'$ must contain either $(x, y)$ or $(y, x)$. Let $S_i'$ be the first occurrence of such a pair.

First suppose $S_i' = (x, y)$. Then for each intermediate set of networks $\mathcal{N}SS'_{[:j]}$ for $j < i$ we have that all the networks in the set either do not contain $x$, or have the reducible pair $(x, y)$. Hence, by repeated application of Lemma 13, there is a sequence $S(x, y)S''$ for $\mathcal{N}$. This sequence has the same length as $SS'$, because it is simply a reordering of the pairs.

Now suppose $S_i' = (y, x)$, then $x$ cannot have been the first coordinate in any pair of $S$, so all networks in $\mathcal{N}S$ contain $x$. Furthermore, $S'$ does not contain the pair $(x, y)$, as this would violate the assumption that $SS'$ is a TCS. Hence, each network in $\mathcal{N}S$ has a cherry or reticulated cherry on $x$ and $y$, which is ultimately reduced by a pair $(y, x)$ in $S'$. Suppose a network $N \in \mathcal{N}S$ does not have the cherry $\{x, y\}$. Then it has the reticulated cherry $(x, y)$. To make this into a cherry, so that it can be reduced by $(y, x)$, the sequence must first contain a pair of the form $(x, z)$. However, this implies $S'$ first contains $(x, z)$ and then $(y, x)$, which contradicts the fact that $SS'$ is a TCS. Hence, we may assume that all networks in $\mathcal{N}S$ have the cherry $\{x, y\}$.

If $y$ is not forbidden after $S$, we can switch the roles of $x$ and $y$ in the remaining part of the sequence $S'$ to get a new TCS $SS^*$ for $\mathcal{N}$. In $S^*$, the first occurrence of $(x, y)$ or $(y, x)$ is $S_i^* = (x, y)$, and we are in the previous case. If $y$ is forbidden after $S$, repeated application of Lemma 13 on $SS'$ and $S_i'$ gives a sequence $S(y, x)S'''$ for $\mathcal{N}$.  □

**Bounding Reducible Pairs in Networks with All Leaves.** In the algorithm in [7], a bound on the number of cherries after having reduced all trivial cherries was required to compute the running time. Here, we require something similar; we require a bound on the number of reducible pairs after we have reduced all the trivial pairs. [7] prove such bounds by first focusing on the case where all input trees have the same leaf set. We do the same, by first focusing on the case where all input networks have the same leaf set.

Let $\mathcal{N}$ be a set of networks. Then the *set of displayed trees of* $\mathcal{N}$ is the set of all trees that are displayed by the networks of $\mathcal{N}$.

**Lemma 15.** *Let* $\mathcal{N} = \{N_1, \ldots, N_I\}$ *be a set of tree-child networks on a common set of leaves such that there exists a TCS S for* $\mathcal{N}$. *If* $\mathcal{N}$ *does not contain any trivial pairs, then the set of displayed trees of* $\mathcal{N}$ *has no trivial cherries.*

**Lemma 16** ([7] **Lemma 10**). *Let* $\mathcal{T}$ *be a set of phylogenetic trees with leaf set* $X$ *such that there is a tree-child network* $N$ *with* $k$ *reticulations that displays* $\mathcal{T}$. *If* $\mathcal{T}$ *has no trivial cherries, then the total number of cherries of the trees in* $\mathcal{T}$ *is at most* $4k$.

Lemmas 15 and 16 gives the bound on the number of reducible pairs for networks with common leaf sets.

**Lemma 17.** *Let* $\mathcal{N}$ *be a set of tree-child networks with leaf set* $X$ *such that there is a tree-child network* $N$ *with* $k$ *reticulations that displays* $\mathcal{N}$. *If* $\mathcal{N}$ *has no trivial pairs, then the total number of reducible pairs of the networks in* $\mathcal{N}$ *is at most* $8k$.

*Proof.* Each reducible pair of a network is a cherry in one of its displayed trees, and the set of displayed trees is displayed by the solution network $N$ as well. Hence, by Lemma 16, there are at most $8k$ reducible pairs in the trees, and therefore at most $8k$ reducible pairs in the networks. □

**Bounding Reducible Pairs in General.** Recall that the algorithm will build a TCS by successively appending reducible pairs; it terminates upon finding the shortest possible sequence that reduces all the input networks. In the process, it branches on all possible non-trivial pairs that the input network may have. Depending on the sequence that is being built, it is possible that leaves that exist in some of the input networks (after reduction by the existing sequence) may have already been deleted from others. Here, we show that even for these instances, it is still the case that the number of possible reducible pairs that we can branch on is bounded by $8k$. This result follows directly from Lemma 7 of [7]: we change the wording of the statement slightly to accommodate for network inputs.

**Lemma 18.** *Let* $\mathcal{N}$ *be a set of tree-child networks on* $X$, *and let* $S = (x_1, y_1)$, $(x_2, y_2), \ldots, (x_r, y_r)$ *be a TCS for* $\mathcal{N}$ *with weight* $k$. *For any* $j \in [r] \cup \{0\}$, *either there exists a trivial pair of* $\mathcal{N}S_{[:j]}$, *or* $\mathcal{N}S_{[:j]}$ *has at most* $8k$ *reducible pairs.*

The idea of the proof is as follows. Let $j$ be such that $\mathcal{N}S_{[:j]}$ has no trivial pairs. Then we find a set of tree-child networks $\hat{\mathcal{N}}_j$ on $X$ with the same set of reducible pairs as $\mathcal{N}S_{[:j]}$ and tree-child hybridization number at most $k$. By Lemma 17, this shows that $\mathcal{N}S_{[:j]}$ has at most $8k$ reducible pairs.

The set of networks is constructed by adding back each missing leaf to each network in $\mathcal{N}S_{[:j]}$ at the root. The order in which they are placed at the root is the same as the order in which these leaves appear as first element in $S_{[:j]}$. Now, we may construct a TCS of the same weight as $S$ that reduces this set of networks. By first reducing the part that corresponds to the part in $\mathcal{N}S_{[:j]}$, and then the leaves placed by the root, we have a TCS that reduces $\hat{\mathcal{N}}_j$ of weight at most $k$:

$$(x_{j+1}, y_{j+1}), (x_{j+2}, y_{j+2}), \ldots, (x_r, y_r), (x_1, y_r), (x_2, y_r), \ldots, (x_j, y_r).$$

An example of the corresponding networks and their embeddings can be found in Fig. 3.

$S = (4,3)(1,2) \circ (4,5)(3,2)(3,5)(1,2)(2,5)$ $\qquad$ $\hat{S} = (4,5)(3,2)(3,5)(1,2)(2,5) \circ (4,5)(1,5)$

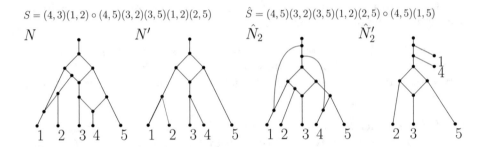

**Fig. 3.** A set $\mathcal{N} = \{N, N'\}$ of tree-child networks on the set of taxa $\{1, 2, 3, 4, 5\}$, with a TCS $S$ that reduces it. A set $\hat{\mathcal{N}}_2 = \{\hat{N}_2, \hat{N}_2'\}$ of tree-child networks obtained by reducing the first two elements of $S$ from $\mathcal{N}$, and reattaching the tail of the deleted edges (red edge) to the root edge, in the order that they were deleted in (as explained in the sketch proof of Lemma 18). The sequence $\hat{S}$ is a TCS of the same weight as $S$, obtained from $S$ by deleting the first two elements and appending these two elements to the end of the sequence, for which we replace the second coordinate of the elements by 5 (the leaf that appears as the second coordinate element in the last element of $S$.

## 4.2   Adapting the Algorithm

Our algorithms are the same as those presented in [7], except for the following changes.

- The input set of trees $\mathcal{T}$ is changed into an input set of tree-child networks $\mathcal{N}$;
- trivial cherries are now trivial pairs;
- In line 4, the stop condition of a non-pickable reticulated cherry is added;

The first change is necessary for the algorithm to take an input consisting of networks. The second change is necessary as not all reducible pairs are cherries anymore, when the input consists of networks. The while-loop that reduces all the trivial pairs is still correct in the algorithm, because there is an optimal sequence that first reduces all trivial pairs (Lemma 14). The last change makes

---

**Procedure** TreeChildSequence($\mathcal{N}, S, k$)

---

**Input**: A collection $\mathcal{N}$ of tree-child networks, a partial TCS $S$, an integer $k$;
**Output**: An optimal TCS $SS'$ of weight at most $k$ for $\mathcal{N}$ if such a sequence exists; FAIL otherwise;

1  **while** *There exists a trivial pair $(x, y)$ in $\mathcal{N}S$ with $y$ not forbidden by $S$* **do**
2      Set $S = S(x, y)$;

3  Set $\mathcal{N}' = \mathcal{N}S$;
4  **if** *some network in $\mathcal{N}'$ has a cherry $(x, y)$ with $x, y$ forbidden or a reticulated cherry $(x, y)$ with $y$ forbidden* **then**
5      **return** FAIL;

6  **else**
7      Set $n' = |\{x \in X : x$ is a leaf in $\mathcal{N}'\}|$;
8      Set $k' = |S| - |X| + n'$;
9      Set $C = \{(x, y) \mid (x, y)$ is a reducible pair of some network in $\mathcal{N}'\}$;
10     **if** $|C| == 0$ **then**
11        **return** $S$;
12     **else if** $|C| > 8k$ *or* $k' \geq k$ **then**
13        **return** FAIL;
14     **else**
15        Set $S_{opt} = $ FAIL;
16        **foreach** $(x, y) \in C$ *with $y$ not forbidden by $S$* **do**
17           Set $S_{temp} = $ TreeChildSequence($\mathcal{N}, S(x, y), k$);
18           **if** $S_{temp} \neq$ FAIL *and* $(S_{opt} = $ FAIL *or* $(S_{opt} \neq$ FAIL *and* $w(S_{temp}) < w(S_{opt})))$ **then**
19              Set $S_{opt} = S_{temp}$;

20        **return** $S_{opt}$;

---

sure we stop when the reduced input $\mathcal{N}S$ cannot be fully reduced using a TCS that can be appended after the prefix $S$.

Otherwise, the algorithm is still correct. Indeed, the algorithm branches over all non-trivial pairs, to find a shortest sequence that reduces all input networks; and this shortest sequence corresponds to a network with minimal reticulation number that displays all input networks. Furthermore, the running time follows as each branch-out is over at most $8k$ pairs, and the search depth is at most $k$.

**Theorem 19.** *Let $\mathcal{N}$ be a set of tree-child networks on a set of taxa $X$. If there exists a tree-child network with at most $k$ reticulations that displays $\mathcal{N}$, then it can be found in $O((8k)^k \cdot \mathrm{poly}(|X|, |\mathcal{N}|))$ time using* TreeChildNetwork($\mathcal{N}, k$).

## 5 Discussion

In this paper, we have introduced NETWORK HYBRIDIZATION, the problem of finding a network with minimal reticulation number that displays a set of net-

---

**Procedure** TreeChildNetwork($\mathcal{N}, k$)

---

**Input**: A collection $\mathcal{N}$ of tree-child networks, an integer $k$;
**Output**: A tree-child phylogenetic network $N$ on $X$ that displays $\mathcal{N}$ with
        reticulation number at most $k$, if such a network exists; otherwise
        NONE;

1  Set $S =$ TreeChildSequence($\mathcal{N}, \emptyset, k$);
2  if $S ==$ FAIL **then**
3    |  **return** NONE;

4  **else**
5    |  Set $N =$ ConstructNetworkFromSequence($S$);
6    |  **return** $N$;

---

works. We showed that the TREE-CHILD NETWORK HYBRIDIZATION problem, in which we restrict our inputs and output to be tree-child networks, can be solved by making slight adjustments to the FPT algorithm presented in [7].

In practice, our algorithm can be sped up using the heuristic improvement that was introduced in [7]. We may consider branch reduction, in which we ignore parts of the search tree where no better solution can be found.

For this problem, FPT is essentially the best we can do, because solving the NETWORK HYBRIDIZATION problem for an input set of tree-child networks is NP-hard. This follows from the fact that it is already NP-hard for an input set of trees. It has recently been shown that if all level-$(k-1)$ subnetworks of a level-$k$ tree-child networks are given, this network can be constructed in polynomial time [15]. In other words, the TREE-CHILD NETWORK HYBRIDIZATION problem is easy to solve when we are given all level-$(k-1)$ subnetworks of a level-$k$ network. This suggests that the problem becomes easy if the difference in reticulation number between the inputs and the output network is bounded. We wonder if this is still true for networks that are not tree-child, and therefore it would be interesting to see whether the HYBRIDIZATION problem is FPT with this difference in reticulation number as parameter. And, if this is the case, whether the current algorithm can be proven to have this running time.

Recall that a TCS is a sequence of ordered pairs with two conditions imposed on them: the first condition ensures that we obtain a network from the sequence upon using the CONSTRUCTNETWORKFROMSEQUENCE algorithm; the second condition ensures that the network we obtain is tree-child. Upon removing this second condition from sequences of ordered pairs, we obtain what is called a *cherry-picking sequence* [13]. Networks that can be reduced by a cherry-picking sequence are called *orchard* networks [3,13]. A natural extension of the results we have presented in this paper would be to consider the following problem.

---

ORCHARD NETWORK HYBRIDIZATION
**Input:** A set of orchard networks $\mathcal{N}$ on $X$.
**Output:** An orchard network that displays $\mathcal{N}$ with minimal reticulation number.

---

Ideally, in Algorithm TREECHILDSEQUENCE, we would simply remove the tree-child condition to obtain an algorithm which works for orchard networks as well. However, simply doing so could potentially result in a much higher running time, as we do not have a bound on the number of reducible pairs for orchard networks (see Fig. 4). Nevertheless, solving ORCHARD NETWORK HYBRIDIZATION could lead to better upper bounds for the network hybridization number, and the algorithm could still be efficient in practice. In this light, this paper has taken the first step towards finding good solutions for NETWORK HYBRIDIZATION.

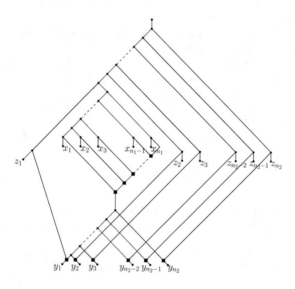

**Fig. 4.** An orchard network with $n_1 + n_2 - 1$ reticulations such that the set of displayed trees have at least $n_1 n_2$ cherries.

# References

1. Baroni, M., Semple, C., Steel, M.: A framework for representing reticulate evolution. Ann. Comb. **8**(4), 391–408 (2005)
2. Bordewich, M., Semple, C.: Computing the minimum number of hybridization events for a consistent evolutionary history. Discret. Appl. Math. **155**(8), 914–928 (2007)
3. Erdős, P.L., Semple, C., Steel, M.: A class of phylogenetic networks reconstructable from ancestral profiles. Math. Biosci. **313**, 33–40 (2019)
4. Huber, K.T., Moulton, V., Semple, C., Wu, T.: Quarnet inference rules for level-1 networks. Bull. Math. Biol. **80**(8), 2137–2153 (2018)
5. Huber, K.T., van Iersel, L., Moulton, V., Scornavacca, C., Wu, T.: Reconstructing phylogenetic level-1 networks from nondense binet and trinet sets. Algorithmica **77**(1), 173–200 (2017)

6. Humphries, P.J., Linz, S., Semple, C.: Cherry picking: a characterization of the temporal hybridization number for a set of phylogenies. Bull. Math. Biol. **75**(10), 1879–1890 (2013)

7. van Iersel, L., Janssen, R., Jones, M., Murakami, Y., Zeh, N.: A practical fixed-parameter algorithm for constructing tree-child networks from multiple binary trees. arXiv preprint arXiv:1907.08474 (2019)

8. van Iersel, L., Kelk, S., Lekic, N., Whidden, C., Zeh, N.: Hybridization number on three rooted binary trees is EPT. SIAM J. Discret. Math. **30**(3), 1607–1631 (2016)

9. van Iersel, L., Linz, S.: A quadratic kernel for computing the hybridization number of multiple trees. Inf. Process. Lett. **113**(9), 318–323 (2013)

10. van Iersel, L., Moulton, V.: Trinets encode tree-child and level-2 phylogenetic networks. J. Math. Biol. **68**(7), 1707–1729 (2014)

11. van Iersel, L., Moulton, V.: Leaf-reconstructibility of phylogenetic networks. SIAM J. Discret. Math. **32**(3), 2047–2066 (2018)

12. van Iersel, L., Moulton, V., de Swart, E., Wu, T.: Binets: fundamental building blocks for phylogenetic networks. Bull. Math. Biol. **79**(5), 1135–1154 (2017)

13. Janssen, R., Murakami, Y.: Solving phylogenetic network containment problems using cherry-picking sequences. arXiv preprint arXiv:1812.08065 (2018)

14. Linz, S., Semple, C.: Attaching leaves and picking cherries to characterise the hybridisation number for a set of phylogenies. Adv. Appl. Math. **105**, 102–129 (2019)

15. Murakami, Y., van Iersel, L., Janssen, R., Jones, M., Moulton, V.: Reconstructing tree-child networks from reticulate-edge-deleted subnetworks. Bull. Math. Biol. **81**(10), 3823–3863 (2019)

16. Oldman, J., Wu, T., Van Iersel, L., Moulton, V.: Trilonet: piecing together small networks to reconstruct reticulate evolutionary histories. Mol. Biol. Evol. **33**(8), 2151–2162 (2016)

17. Whidden, C., Beiko, R.G., Zeh, N.: Fixed-parameter algorithms for maximum agreement forests. SIAM J. Comput. **42**(4), 1431–1466 (2013)

# Linear Time Algorithm for Tree-Child Network Containment

Remie Janssen$^{(\boxtimes)}$ and Yukihiro Murakami

Delft Institute of Applied Mathematics, Delft University of Technology,
Van Mourik Broekmanweg 6, 2628 XE Delft, The Netherlands
{R.Janssen-2,Y.Murakami}@tudelft.nl

**Abstract.** Phylogenetic networks are used to represent evolutionary
scenarios in biology and linguistics. To find the most probable scenario, it
may be necessary to compare candidate networks, to distinguish different
networks, and to see when one network is embedded in another. Here,
we consider the NETWORK CONTAINMENT problem, which asks whether
a given network is contained in another network. We give a linear-time
algorithm to this problem for the class of tree-child networks using the
recently introduced tree-child sequences by Linz and Semple. We implement this algorithm in Python and show that the linear-time theoretical
bound on the input size is achievable in practice.

**Keywords:** Phylogenetics · Tree-child networks · Network
Containment · Tree-child sequences

## 1 Introduction

Phylogenetic networks are gaining popularity in the study of the evolutionary
history of taxa [1,12]. However, small stretches of DNA (e.g., pieces of DNA
coding for protein domains) evolve tree-like. Therefore, the network representing
the species' evolution must contain the trees for such pieces of DNA. This leads
to the following mathematical problem. For a given network $N$ and a tree $T$ on
the same set of taxa, decide whether $N$ contains $T$.

This problem, called TREE CONTAINMENT, is NP-complete for general rooted
phylogenetic networks [10]. The problem remains NP-complete for certain network classes (networks with particular topological restrictions), such as tree-
sibling, time-consistent, and regular networks [8]. However, for other network
classes, the problem becomes easier. For example, it is known that TREE CON-
TAINMENT can be solved in polynomial time for normal networks, tree-child
networks, and level-$k$ networks [8].

There are even stronger results for some network classes: deciding whether a
tree is contained in a genetically stable network can be done in quadratic time

---

Research funded by the Netherlands Organization for Scientific Research (NWO), with
the Vidi grant 639.072.602.

C. Martín-Vide et al. (Eds.): AlCoB 2020, LNBI 12099, pp. 93–107, 2020.
https://doi.org/10.1007/978-3-030-42266-0_8

[5], and making this decision for a binary nearly-stable network takes linear time [6]. For the class of tree-child networks, TREE CONTAINMENT is known to be solvable in linear time [6,7].

From a biological and a computational perspective, there is no reason why we should restrict ourselves to inputs of a tree and a network. Indeed, while small stretches of DNA may evolve tree-like, it is possible for another part of the genome to evolve as a network. In such instances, it is of great interest to consider a more general version of TREE CONTAINMENT, which we call NETWORK CONTAINMENT: For given networks $N$ and $N'$ on the same set of taxa, decide whether $N$ contains $N'$. By extension, the problem remains NP-complete for inputs of general rooted phylogenetic networks. Computationally, it is natural to wonder whether network classes that can solve TREE CONTAINMENT efficiently can also solve NETWORK CONTAINMENT in a similar fashion. To date, no study has ever considered this problem, and we take the first steps in this endeavour.

We solve the NETWORK CONTAINMENT problem for tree-child networks (defined formally in Sect. 2) by considering *tree-child sequences*. These sequences were developed to tackle the problem of finding a "simple" network that contains a given set of trees [4,11]. Two leaves of a tree form a *cherry* if they share a common parent—by successively *picking* cherries (removing one of the leaves in a cherry) from the set of input trees, we obtain a sequence of cherries that ultimately reduce each input tree to a tree on a single leaf. This sequence of cherries then corresponds to some network that contains the set of all input trees.

Previously, these reductions were only defined on trees, and not on networks. In this paper, we start by defining tree-child sequences and their actions on tree-child networks. We show that for every tree-child network, there exists a sequence of ordered pairs of leaves that reduces it to a network on a single leaf. The order in which these pairs are picked does not matter. We also show that a tree-child network is contained in another tree-child network if and only if a sequence that reduces the first network also reduces the second one. Combining these results culminates in a linear-time algorithm for solving the NETWORK CONTAINMENT problem for tree-child networks.

*Structure of the Paper.* We start by giving all relevant definitions and outlining how to construct networks from tree-child sequences (Sect. 2). In Sect. 3, we investigate properties of tree-child sequences, and their relation to tree-child networks. In particular, we focus on the relation between tree-child subsequences, and subnetworks of tree-child networks. This section also includes an algorithm to solve TREE-CHILD NETWORK CONTAINMENT. Then, in Sect. 4, we present an efficient implementation of this algorithm in Python, and show that the theoretical running time is achievable in practice. We test our implementation on simulated data, and show that even for large data sets (1000 leaves and 1000 reticulations), the software outputs the solution within a tenth of a second. Lastly, in Sect. 5, we conclude with open problems and future directions for the use of cherry-picking strategies.

# 2    Preliminaries

**Definition 1.** *A* phylogenetic semi-binary network *N* is a DAG with one *outdegree-1 source (the root), a set $L(N)$ of indegree-1 sinks (leaves) bijectively labelled with a set $X$, and all other nodes are either of indegree-1 and outdegree-2 (tree nodes) or of indegree at least 2 and outdegree-1 (reticulations).*

A network is *binary* if all reticulations have indegree-2. In this paper, all networks we consider are phylogenetic semi-binary networks unless stated otherwise, so we call these networks for short. Furthermore, all networks have the leaf set $X = \{1, 2, \ldots, n\}$, unless stated otherwise.

An edge feeding into reticulations is called a *reticulation edge*. Given an edge $uv$ in $N$, we say that $u$ is a *parent* of $v$ and that $v$ is a *child* of $u$. The node $u$ is *above* $v$ if there is a directed path from $u$ to $v$ in $N$. The network $N$ is *tree-child* if every non-leaf node in $N$ is a parent of a tree node or a leaf. The *reticulation number* is the total number of reticulation edges minus the total number of reticulations.

Let $N$ and $N'$ be tree-child networks on the same set of taxa $X$. Then $N$ *contains* $N'$ if $N'$ can be obtained from $N$ by deleting reticulation edges and suppressing degree-2 nodes. We now formally define the TREE-CHILD NETWORK CONTAINMENT problem.

---

TREE-CHILD NETWORK CONTAINMENT
**Instance:** Two tree-child networks $N$ and $N'$ on the same leaf-set.
**Question:** Does $N$ contain $N'$?

---

## 2.1    Reducible Pairs

Let $(x, y)$ be an ordered pair of leaves in a network $N$, and let $p_x, p_y$ denote the parents of $x, y$ respectively. We call $(x, y)$ a *cherry* if $p_x = p_y$, if $x$ and $y$ share a common parent. Observe that if $(x, y)$ is a cherry, then $(y, x)$ must also be a cherry. We call $(x, y)$ a *reticulated cherry* if $p_x$ is a reticulation and $p_y$ is a parent of $p_x$. If $(x, y)$ is a cherry or a reticulated cherry in $N$, we call this a *reducible pair*. The following algorithms show that finding reducible pairs of a network can be done quickly. Observe that since tree nodes are of outdegree-2, each leaf appears as a second coordinate in at most one reducible pair in a network; Algorithm 1 finds such a reducible pair, if it exists, for a given leaf in constant time. Algorithm 2 on the other hand finds all reticulated cherries that contain a given leaf as the first coordinate of the reducible pair. The running time for this algorithm depends on the indegree of the parent of the given leaf, as this gives the maximum possible number of such reticulated cherries.

---

**Algorithm 1.** FINDRP2ND($N, x$)

---

**Data:** A network $N$ and a leaf $x$

**Result:** The set containing exactly one reducible pair of $N$ that has $x$ as the second coordinate if it exists; $\emptyset$ otherwise.

1 Let $p$ be the parent of $x$;
2 **if** $p$ *is a tree node* **then**
3      let $c(p)$ be the child of $p$ that is not $x$;
4      **if** $c(p)$ *is a leaf* **then**
5          **return** $\{(c(p), x)\}$;
6      **if** $c(p)$ *is a reticulation and the child* $c(c(p))$ *of* $c(p)$ *is a leaf* **then**
7          **return** $\{(c(c(p)), x)\}$;
8      **end**
9 **end**
10 **return** $\emptyset$;

---

**Lemma 2.** *Let $x$ be a leaf in a network $N$. If a reducible pair with $x$ as the second element of the pair exists, then Algorithm 1 finds this pair in constant time. Otherwise it returns the empty set in constant time.*

---

**Algorithm 2.** FINDRC1ST($N, x$)

---

**Data:** A network $N$ and a leaf $x$

**Result:** The set of all reticulated cherries in $N$ that has $x$ as the first coordinate

1 Let $p$ be the parent of $x$;
2 Set $C_r = \emptyset$;
3 **if** $p$ *is a reticulation* **then**
4      **for** *every parent $g$ of $p$* **do**
5          let $c(g)$ be the child of $g$ that is not $p$;
6          **if** $c(g)$ *is a leaf* **then**
7              $C_r = C_r \cup \{(x, c(g))\}$
8      **end**
9 **end**
10 **return** $C_r$;

---

**Lemma 3.** *Let $x$ be a leaf in a network $N$, and let $p_x$ denote the parent of $x$. Let $I$ denote the indegree of $p_x$. Algorithm 2 finds the set of all reticulated cherries that has $x$ as the first coordinate in $O(I)$ time.*

## 2.2   Reducing Pairs from Networks

Given a cherry or a reticulated cherry, we may *reduce* them from a network to obtain a network of smaller size.

**Definition 4.** *Let $N$ be a network and let $(x, y)$ be an ordered pair of leaves. Reducing $(x, y)$ in $N$ is the action of*

- *deleting $x$ and suppressing its parent node in $N$ if $(x, y)$ is a cherry in $N$;*
- *deleting the reticulation edge between the parents of $x$ and $y$ and subsequently suppressing degree-2 nodes, if $(x, y)$ is a reticulated cherry;*
- *doing nothing to $N$ otherwise.*

*In all cases, the resulting network is denoted $N(x, y)$.*

We refer to this as *picking a reducible pair $(x, y)$ from $N$*. We transform this definition into an algorithm, and show that a reduction of a pair from a network can be done in constant time.

---

**Algorithm 3.** REDUCEPAIR($N, (x, y)$)

---

**Data:** A network $N$ and a pair of leaves $(x, y)$
**Result:** The network $N(x, y)$
1 **if** $(x, y)$ *is a cherry in $N$* **then**
2    | Let $p$ be the parent of $x$ and $y$;
3    | Remove edge $px$ from $N$;
4    | Suppress $p$ (if it is a node of degree-2) and remove $x$ in $N$;
5 **if** $(x, y)$ *is a reticulated cherry in $N$* **then**
6    | Let $p_x$ be the parent of $x$ and $p_y$ the parent of $y$;
7    | Remove edge $p_y p_x$ from $N$;
8    | Suppress $p_y$ and $p_x$ (if they are nodes of degree-2) in $N$;
9 **end**
10 **return** $N$;

---

**Lemma 5.** *Algorithm 3 correctly reduces a given reducible pair in a network $N$ in constant time.*

## 3   Tree-Child Sequence

In this section, we formally define *tree-child sequences* and how they correspond to tree-child networks. These are sequences of ordered pairs with additional properties; to illustrate the intuition behind these properties, we start by showing how to construct networks from sequences of ordered pairs.

**Definition 6.** *Let $N$ be a network and let $(x, y)$ be reducible pair. Then we may construct $N$ from $N(x, y)$—also called* add $(x, y)$ to $N(x, y)$—*by applying the following.*

1. *If $x$ is a leaf in $N(x, y)$ (i.e., if $(x, y)$ is a reticulated cherry in $N$), and*
   *(a) if $p$, the parent of $x$ in $N(x, y)$, is a reticulation then add a node $q$ directly above $y$, and add an edge $qp$.*
   *(b) otherwise, add nodes $p$ and $q$ directly above $x$ and $y$ respectively, and add an edge $qp$.*

2. *If $x$ is not a leaf in $N(x, y)$ (i.e., if $(x, y)$ is a cherry in $N$) then add a labelled node $x$, insert a node $q$ directly above $y$, and add an edge $qx$.*

Observe that when adding $(x, y)$ to $N(x, y)$, we assume that $y$ is a leaf in the network $N(x, y)$. Otherwise, adding $(x, y)$ to $N(x, y)$ is not well-defined.

Now let $S = S_1 S_2 \cdots S_{|S|} = (x_1, y_1)(x_2, y_2) \cdots (x_{|S|}, y_{|S|})$ be a sequence of ordered pairs with the condition that the second coordinate of each pair occurs as a first coordinate in the rest of the sequence, or as the second coordinate of the last pair. Starting with a network on a single leaf $y_{|S|}$, we may iteratively add $S_i$ to the network for $i = |S|, |S| - 1, \ldots, 1$ (i.e., backwards through the sequence $S$) to obtain some network. We call this *the network obtained from $S$*. This condition ensures that when adding $(x_i, y_i)$ to the network, $y_i$ is already a leaf in the network so that the operation is well-defined.

Now suppose that we add a second condition on $S$ that the first coordinate of each pair does not appear as a second coordinate of another pair in the remainder of the sequence. We will sometimes refer to this condition as the *tree-child condition*. Then, we claim that the network obtained from $S$ is tree-child. By construction, we never obtain reticulation nodes that are adjacent to one another. In particular, every reticulation edge is inserted to existing reticulation nodes whenever possible (Definition 6.1a). Hence, we may only violate the tree-child property from a tree node having two reticulation children. So say that we have just added a reticulated cherry $(x, y)$ to a network $N$. In $N$, the tree node parent $p_y$ of $y$ currently has one reticulation child and one leaf child $y$. For $p_y$ to have two reticulation children, we require some reticulation node to be inserted between $p_y$ and $y$, which can only happen if we add some ordered pair $(y, z)$ to $N$. However, this would mean that $y$ appears as a first coordinate of some pair and also as a second coordinate of some pair later on in the sequence, which contradicts our second condition. If, on the other hand, we have just added a cherry $(x, y)$ to $N$, then the parent $p$ of $x$ cannot be a parent of two reticulations after adding more reducible pairs. Indeed, this would imply that we have added some reducible pair $(y, z)$ later on to the network (and hence it would appear earlier in the sequence), which again contradicts our second condition.

This brings us to the following definition.

**Definition 7.** *A tree-child sequence (TCS) is a sequence of ordered pairs of two leaves such that*

- *the second coordinate of each pair occurs as a first coordinate in the rest of the sequence, or as the second coordinate of the last pair; and*
- *no first coordinate leaf is used as a second coordinate in the remainder of the sequence.*

Let $N$ be a network and let $S$ be a TCS. Denote by $NS$ the network obtained by repeatedly reducing $N$ with each element of $S$ in order. We say that $S$ *reduces $N$* if $NS$ is a network with a single leaf (for any leaf in $N$), a root, and no other nodes. We call a TCS $S$ *minimal* for a tree-child network $N$ if $S$ reduces $N$ and if $NS_1 \cdots S_{i-1} \neq NS_1 \cdots S_i$ for all $i \in [|S|]$. Suppose that $N$ contains $n$ leaves and has reticulation number $r$. Then any minimal TCS for $N$ is of length $n + r - 1$.

Using the operations outlined in Definition 6, one may obtain a tree-child network $N$ from a given TCS $S$. As each addition of an ordered pair creates either a cherry or a reticulated cherry in the network, we may simply reverse the operations to see that $S$ reduces $N$. This brings us to the following correspondence.

**Theorem 8.** *Let $N$ be a tree-child network. Then there exists a minimal TCS $S$ that reduces it. The network obtained from $S$ is isomorphic to $N$.*

*Let $S$ be a TCS. Then the network obtained from $S$ is unique and is tree-child. Furthermore, $S$ is a minimal TCS for this network.*

While each TCS gives rise to a unique tree-child network, there can be many TCSs that reduce the same tree-child network. In particular, given a tree-child network, we may pick the reducible pairs in any order.

**Theorem 9.** *Let $N$ be a tree-child network and let $(x, y)$ be a reducible pair of $N$. Then there exists a minimal TCS of $N$ whose first element is $(x, y)$.*

In the setting of Theorem 9, we have that $N(x, y)$ is a tree-child network. Then, by iteratively applying the theorem to the reduced network each time, it is indeed the case that we may pick the reducible pairs in any order—making sure the second property of a TCS is not violated. The following algorithm then shows how we may obtain a minimal TCS for a tree-child network by picking reducible pairs in any order, and maintaining a list of all reducible pairs in the network.

---

**Algorithm 4.** FINDTCS($N$)

---

**Data:** A tree-child network $N$

**Result:** A minimal TCS $S$ for $N$

1  Set $\mathcal{C} = \emptyset$ ;
2  **for** $x \in L(N)$ **do**
3  $\quad$ $\mathcal{C} \cup$ FINDRP2ND($N, x$);
4  **end**
5  Let $S$ be an empty sequence;
6  **while** $\mathcal{C} \neq \emptyset$ **do**
7  $\quad$ Choose $(x, y) \in \mathcal{C}$;
8  $\quad$ Set $S = S(x, y)$;
9  $\quad$ $N' =$ REDUCEPAIR($N, (x, y)$);
10 $\quad$ **if** $(x, y)$ *is a cherry in $N$* **then**
11 $\quad\quad$ $\mathcal{C} = \mathcal{C} \setminus \{(x, y), (y, x)\} \cup$ FINDRP2ND($N', y$)$\cup$FINDRC1ST($N', y$);
12 $\quad$ **if** $(x, y)$ *is a reticulated cherry in $N$* **then**
13 $\quad\quad$ $\mathcal{C} = \mathcal{C} \setminus \{(x, y)\} \cup$ FINDRP2ND($N', y$)$\cup$FINDRC1ST($N', y$);
14 $\quad$ **end**
15 $\quad$ $N = N'$;
16 **end**
17 **return** $S$;

---

**Lemma 10.** *Let $N$ be a tree-child network on $X$ with reticulation number $r$. Algorithm 4 finds a minimal TCS for $N$ in $O(n + r)$ time.*

### 3.1   Putting It All Together

The following theorem characterizes when a tree-child network is contained in another tree-child network, using TCSs.

**Theorem 11.** *Let $N$ and $N'$ be two tree-child networks on the same leaf-sets. $N$ contains $N'$ if and only if any minimal TCS of $N$ reduces $N'$.*

Therefore, using the subroutines that we have introduced previously (Algorithms 1–4), we obtain the following algorithm that solves the TREE-CHILD NETWORK CONTAINMENT problem. Let $N$ and $N'$ be two tree-child networks on the same leaf-sets. Using Theorem 9, we first obtain some minimal sequence $S$ that reduces $N$ by picking reducible pairs in any order (Algorithm 4). By Theorem 11, if $S$ reduces $N'$, then $N'$ is contained in $N$; otherwise, $N'$ is not contained in $N$.

---

**Algorithm 5.** TCNCONTAINS$(N, N')$

---

**Data:** Two tree-child networks $N$ and $N'$ on the same set of taxa
**Result:** Yes if $N$ contains $N'$, No otherwise.
1  Set $S =$ FINDTCS$(N)$;
2  **for** $i = 1, \ldots, |S|$ **do**
3  |    $N' =$ REDUCEPAIR$(N', S_i)$;
4  **end**
5  **if** $N'$ *is a network on a single leaf* **then**
6  |    **return** Yes;
7  **end**
8  **return** No;

---

**Theorem 12.** *Given two tree-child networks $N$ and $N'$ on the same taxa set $X$ where the reticulation number of $N$ is $r$, it can be decided in time $O(n + r)$ whether $N'$ is contained in $N$.*

The theorem has the following corollary regarding the NETWORK ISOMORPHISM problem, which asks whether two given networks are isomorphic. Indeed, we can solve this problem by running Algorithm 5 twice, since two networks are isomorphic if and only if they are contained in one another. The problem for tree-child networks was previously shown to be solvable in $O(n^2)$ time [3]. Therefore, we present the first linear-time algorithm for checking whether two tree-child networks are isomorphic.

**Corollary 13.** *Given two tree-child networks $N$ and $N'$ on taxa set $X$ where the reticulation number of $N$ is $r$, it can be decided in $O(n + r)$ time whether $N$ is isomorphic to $N'$ (Fig. 1).*

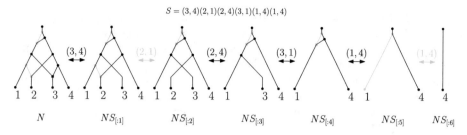

$$S = (3,4)(2,1)(2,4)(3,1)(1,4)(1,4)$$

**Fig. 1.** A binary tree-child network $N$ (grey and black) reduced to a leaf 4 by a tree-child sequence $S$. The reduction is shown as a sequence of networks $NS_{[:i]}$ for $i = 0, 1, \ldots, 6$ from left to right, in which an element of $S$ is applied to the network successively. This sequence is minimal for the network, as every element of the sequence reduces either a cherry or a reticulated cherry of the network. An example of a cherry $(3,1)$ can be seen in the network $NS_{[:3]}$, and a reticulated cherry $(3,4)$ can be seen in the network $N$. The reduction of both reducible pairs is carried out as in Subsect. 2.1. Observe that this sequence is a tree-child sequence. The black subnetwork is also reduced by $S$, and the embedding can be constructed by building both networks simultaneously and keeping track of the edges added by the pairs that change the subnetwork (black pairs and arrows).

## 4   Implementation

Algorithm 5, which checks whether a given tree-child network is a subnetwork of another, was implemented in Python to test the theoretical linear bound in practice. In this section, we present running time results of the implementation on a large randomly generated data set. We show that the theoretically proven linear running time is indeed achievable in practice. The tests were run on a Linux system with a quad-core Intel Xeon W3570 running at 1.7 GHz and 24 GB of DDR3 RAM clocked at 1333 MHz. The operating system was Debian GNU/Linux 9 with a 4.19.46-64 Linux kernel. The software was written in Python version 3.7.3.

### 4.1   Generating the Datasets

For the test data, we generated 131200 instances of the TREE-CHILD NETWORK CONTAINMENT problem: two yes-instances and two no-instances for all $n, r, r' \in \{25, 50, \ldots, 975, 1000\}$ with $r' \leq r$, where $n$ is the number of leaves of both networks, $r$ is the reticulation number in the first network, and $r'$ the reticulation number in the second network. Each instance consists of two semi-binary tree-child networks on the same leaf-set, for which we asked whether the first network contained the second network.

For each instance, we generated the first network with $n$ leaves and reticulation number $r$ using Algorithm 6. The second network was generated depending on whether it was a yes- or a no-instance. If it was a yes-instance, a subnetwork with reticulation number $r'$ was obtained using Algorithm 7; for a no-instance, a network on the same number of leaves and reticulation number $r'$ was randomly generated with the same process as the first network (using Algorithm 6).

This way, each generated yes-instance is always a yes-instance for the TREE-CHILD NETWORK CONTAINMENT problem. For the no-instances, however, the random generation of the second network could also give a subnetwork of the first network, but the probability of that happening is very small, as the number of tree-child networks grows very quickly with the number of leaves and reticulations [2].

The dataset used for the experiment along with the code for generating random datasets, and the actual implementation of Algorithm 5 can be found on https://github.com/RemieJanssen/Cherry-picking_TC_Network_Containment.

**Generating Random Networks.** The tree-child networks were randomly generated as TCSs using Algorithm 6. This algorithm takes two positive integers $n$ and $r$, and outputs a tree-child network with $n$ leaves and reticulation number $r$. It starts with the cherry $(1, 2)$, and successively adds leaves as cherries, and reticulated cherries between two leaves that already exist in the network (respecting the tree-child condition).

In the algorithm, this is achieved by building a tree-child sequence backwards. It chooses to add a reducible pair corresponding to a cherry or reticulated cherry uniformly at random until we have added the required number of leaves and reticulation number. To make sure the sequence is a tree-child sequence, we keep a list $NF$ of taxa that are 'non-forbidden', which, in this case, means that the taxon is not currently the child of a tree node that has a reticulation as the other child (i.e., the leaf has not appeared as a second coordinate element of a reducible pair). If a taxon is in $NF$, it is possible to take this taxon as the first element of a pair appended at the start of the sequence. As a tree-child network always has a cherry or a reticulated cherry, $NF$ is never empty. This implies that the algorithm should never output False, but lines 15 and 16 are kept so that the algorithm can easily be adapted to return only binary tree-child networks. To achieve this, one only has to add the line "$NF = NF \setminus \{\text{first\_element}\}$" between lines 21 and 22 in the algorithm. Finally, the algorithm outputs a TCS, from which we can uniquely construct a TCN.

Note that each tree-child network has positive probability of appearing for this process. In fact, each tree-child sequence ending with $(2, 1)$ has positive probability.

Let us now turn to the procedure to generate a tree-child subnetwork (i.e., generating the second network in a yes-instance). For this purpose, we again work with the representation of the networks as tree-child sequences.

We first select ordered pairs from the sequence of the first network, such that the resulting subsequence corresponds to a tree. This is simply done by selecting a pair with first element $x$ for all $x \in X$ uniformly at random. Because the sequence we started with is a tree-child sequence, the subsequence consisting of the chosen pairs is a tree-child sequence as well: suppose $(x, y)$ and $(y, z)$ are selected. Then $(y, z)$ must appear after $(x, y)$, because otherwise $y$ appears as a first element after it has appeared as a second element in the original sequence.

---

**Algorithm 6.** RANDOMTCS($X, r$)

---

**Data:** A set of taxa $X = \{1, \ldots, n\}$, and a reticulation number $r$.
**Result:** A TCS $S$ on $X$ of length $n + r - 1$.

1  Initialize $Y = \{1, 2\}$ the current set of taxa;
2  Initialize $S = (2, 1)$ the current sequence;
3  Initialize $L = n - 2$;
4  Initialize $R = r$;
5  Initialize $NF = \{2\}$;
6  **while** $L > 0$ *or* $R > 0$ **do**
7     | type_added = None;
8     | **if** $|NF| > 0$ *and* $L > 0$ *and* $R > 0$ **then**
9     |   | With probability $\frac{L}{L+R}$, type_added = L;
10    |   | Otherwise, type_added = R;
11    | **else if** $|NF| > 0$ *and* $R > 0$ **then**
12    |   | type_added = R;
13    | **else if** $L > 0$ **then**
14    |   | type_added = L;
15    | **else**
16    |   | **return** False;
17    | **end**
18    | first_element = None;
19    | second_element = None;
20    | **if** *type_added* $= R$ **then**
21    |   | Set first_element to an element of $NF$ chosen uniformly at random;
22    |   | Set $R = R - 1$;
23    | **else**
24    |   | Set first_element to the first element of $X \setminus Y$;
25    |   | Set $L = L - 1$;
26    |   | Set $Y = Y \cup \{\text{first\_element}\}$;
27    |   | Set $NF = NF \cup \{\text{first\_element}\}$;
28    | **end**
29    | Set second_element to an element of $Y \setminus \{\text{first\_element}\}$ chosen uniformly at random;
30    | $NF = NF \setminus \{\text{second\_element}\}$;
31    | $S = (\text{first\_element}, \text{second\_element})S$;
32 **end**
33 **return** $S$;

---

After selecting the pairs that form a *base tree* of the network (a spanning tree contained by the network), we select $r'$ additional pairs that will form the $r'$ reticulations of the subnetwork. By a similar argument as for the base tree, this subsequence is a tree-child sequence. And as it is reduced by the subsequence, it is also reduced by the sequence of the original network. Hence, the network corresponding to the chosen pairs is a tree-child subnetwork of the original network.

---

**Algorithm 7.** RANDOMSUBTCS($S, r'$)

---

**Data:** A TCS $S$ on $X$ of length $n + r - 1$, and a number $r' \leq r$.

**Result:** A sub-TCS $S'$ of $S$ on $X$ of length $n + r' - 1$

1 Let $S$ be indexed by $\{1, \ldots, |S|\}$;

2 Set $I_{S'} = \emptyset$;

3 **for** $x \in X$ **do**

4      Let $I_x$ be the set of indices of pairs of $S$ with $x$ as first element;

5      Pick $i_x$ uniformly at random from $I_x$;

6      Set $I_{S'} = I_{S'} \cup \{i_x\}$;

7 **end**

8 Randomly add $r'$ elements from $\{1, \ldots, |S|\} \setminus I_{S'}$ to $I_{S'}$;

9 Let $S'$ be the subsequence of $S$ consisting of the elements indexed by $I_{S'}$;

10 **return** $S'$;

---

### 4.2 Results

For all yes-instance tests in which the second network was a subnetwork of the first (i.e., the ones generated by Algorithms 7) and 5 correctly returned a true value. Similarly, for all no-instance tests in which the second network was generated randomly and independently from the first network, Algorithm 5 correctly found that the second network was not a subnetwork of the first. This means that, even though there was a non-zero probability that the second network was a subnetwork, this did not happen in any of the instances. We expected this, as the probability of this happening is extremely small.

Note that the largest test instances (1000 leaves, 1000 reticulations) had a running time of approximately 0.1s. This is expected to scale well for even larger instances, as the linear fit of the data is very good. The $R^2$ values for the fits and the linear dependence of the running time on the number of leaves and reticulations can be found in Table 1. For this fit, we performed a standard linear regression with an intercept of 0 (i.e., forced through the origin), which makes sense because the running time should be zero for an empty instance.

Note that the fits become much better when we split the data in instances where the second network is or is not a subnetwork of the first (i.e., between the yes- and the no- instances), even though the dependence of the running time on the parameters does not change much after this split. The most striking difference we can see in this analysis, is the dependence on the reticulation number of the second network.

As shown in Fig. 2, the no-instances were consistently, and marginally, faster than the yes-instances. For varying leaf numbers, the instances where the second network was not a subnetwork (no-instances), were consistently, but marginally, faster than when the second network was a subnetwork (yes-instances) (Fig. 2, Left). This was similarly true for when we varied the reticulation number $r'$ of the second network. The effect of varying $r'$ on instances for when the second network was not a subnetwork (no-instances) was negligible. This can be seen in the right figure of Fig. 2, but also in Table 1, where the order

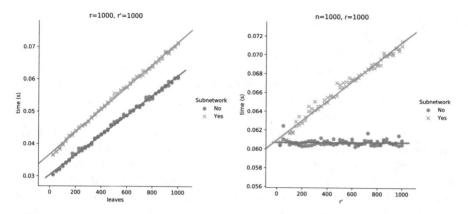

**Fig. 2.** The dependence of the running time on the number of leaves $n$ (left) and the number of reticulations in the second network $r'$ (right). This was visualized by fixing the other parameters to a set value of 1000 in both plots. Fitted lines are independent of Table 1.

of the slope of $r'$ for the no-instances is far smaller than all other slopes in all the instances. For the yes-instances, the running time of the algorithm displayed a linear dependence on $r'$, which was in the same order as the other parameters. This can be explained as follows. When the second network is not a subnetwork, Algorithm 5 will seldom need to reduce a pair in Line 3: it will check whether the pair in the sequence is reducible in the second network. As the second network is randomly generated independently of the first network, it will not have many pairs in common with the first network, which means it will not have to reduce pairs often.

**Table 1.** Linear regression analysis for tree-child network containment on 131200 instances, for which half were yes-instances and the other half no-instances. The high $R^2$ value indicates that the fit of the curve is essentially linear (where an $R^2$ value of 1 indicates a perfect linear fit) and the slopes indicate the change in running time for every increase in the number of leaves, reticulation number $r$, and reticulation number $r'$.

|  | $R^2$ | Slope | | |
|---|---|---|---|---|
|  |  | Leaves (s/leaf) | $r$ (s/reticulation) | $r'$ (s/reticulation) |
| All data | 0.9725659 | $3.03328079 \cdot 10^{-5}$ | $2.99713310 \cdot 10^{-5}$ | $4.75681146 \cdot 10^{-6}$ |
| Subnetwork: YES | 0.9966596 | $3.14405850 \cdot 10^{-5}$ | $2.89850496 \cdot 10^{-5}$ | $9.54505907 \cdot 10^{-6}$ |
| Subnetwork: NO | 0.9976078 | $2.92250310 \cdot 10^{-5}$ | $3.09576119 \cdot 10^{-5}$ | $-3.14361016 \cdot 10^{-8}$ |

## 5  Discussion

In this paper, we have looked at tree-child sequences and how they can be used to solve the NETWORK CONTAINMENT problem for tree-child networks. A theoretical linear-time algorithm was given for this, and we have shown that our

Python implementation also runs in linear time, in the number of leaves and the reticulation number.

In an effort to generalize our results, a natural question would be to ask what would happen if we weakened our current notion of a tree-child sequence. In Definition 7, we stated that the tree-child sequences must satisfy two conditions. The first condition ensures that each sequence corresponds to some network; the second condition ensures that the network is tree-child. Therefore we may consider networks that are more general than tree-child networks, by removing this second condition. These new sequences (which we call *cherry-picking sequences*) may be used to construct networks (called *cherry-picking networks*), with the operations as in Definition 6. This raises two questions. Do our NETWORK CONTAINMENT results hold when we consider inputs of cherry-picking networks? And, is there a structural characterization of these networks? To partly answer these questions for cherry-picking sequences, in [9], we have shown that a network can be reduced in any order (if it can be reduced at all). Furthermore, if a network is reduced by a minimal sequence for another network, the first is contained in the second. However, for cherry-picking sequences, it is no longer true that a subnetwork is reduced by any of the minimal sequences of the original network (Figure 3 of [9]). Therefore, NETWORK CONTAINMENT cannot be solved using cherry-picking sequences; in fact, the counter-example shows that even TREE CONTAINMENT cannot be solved using cherry-picking sequences. The second question, about the structural characterization, remains open.

On a similar note, one can attempt to use tree-child sequences to solve a new problem related to HYBRIDIZATION, where the input is a set of tree-child networks instead of trees. The problem aims to find a tree-child network with minimal reticulation number, containing all input networks. This problem has not been studied before, but could be very important, as there is a dire need of methods for finding a consensus network for a given set of networks.

As a follow-up, it would be interesting to extend our NETWORK CONTAINMENT results to a more general framework. In this paper, we have presented a linear time algorithm for checking whether a tree-child network contains another tree-child network on the same set of taxa. What is the change in complexity (if there is one) when we consider tree-child networks on different sets of taxa? Does the problem become NP-hard, or does it remain polynomial time? Could a modified version of our algorithm be used to solve this problem?

**Acknowledgments.** We would like to thank Leo van Iersel for providing feedback on multiple versions of our paper. We would like to thank Robbert Huijsman for implementing and testing our pseudocode in Python.

# References

1. Bapteste, E., et al.: Networks: expanding evolutionary thinking. Trends Genet. **29**(8), 439–441 (2013)
2. Cardona, G., Pons, J.C., Scornavacca, C.: Generation of binary tree-child phylogenetic networks. PLoS Comput. Biol. **15**(9), e1007347 (2019)

3. Cardona, G., Rossello, F., Valiente, G.: Comparison of tree-child phylogenetic networks. IEEE/ACM Trans. Comput. Biol. Bioinf. **6**(4), 552–569 (2009)
4. Döcker, J., van Iersel, L., Kelk, S., Linz, S.: Deciding the existence of a cherry-picking sequence is hard on two trees. Discrete Appl. Math. **260**, 131–143 (2019)
5. Gambette, P., Gunawan, A.D.M., Labarre, A., Vialette, S., Zhang, L.: Solving the tree containment problem for genetically stable networks in quadratic time. In: Lipták, Z., Smyth, W.F. (eds.) IWOCA 2015. LNCS, vol. 9538, pp. 197–208. Springer, Cham (2016). https://doi.org/10.1007/978-3-319-29516-9_17
6. Gambette, P., Gunawan, A.D., Labarre, A., Vialette, S., Zhang, L.: Solving the tree containment problem in linear time for nearly stable phylogenetic networks. Discrete Appl. Math. **246**, 62–79 (2018)
7. Gunawan, A.: Solving tree containment problem for reticulation-visible networks with optimal running time. arXiv preprint arXiv:1702.04088 (2017)
8. Iersel, L.V., Semple, C., Steel, M.: Locating a tree in a phylogenetic network. Inf. Process. Lett. **110**(23), 1037–1043 (2010)
9. Janssen, R., Murakami, Y.: Solving phylogenetic network containment problems using cherry-picking sequences. arXiv preprint arXiv:1812.08065 (2018)
10. Kanj, I.A., Nakhleh, L., Than, C., Xia, G.: Seeing the trees and their branches in the network is hard. Theoret. Comput. Sci. **401**(1–3), 153–164 (2008)
11. Linz, S., Semple, C.: Attaching leaves and picking cherries to characterise the hybridisation number for a set of phylogenies. Adv. Appl. Math. **105**, 102–129 (2019)
12. Morrison, D.A.: Networks in phylogenetic analysis: new tools for population biology. Int. J. Parasitol. **35**(5), 567–582 (2005)

# PathOGiST: A Novel Method for Clustering Pathogen Isolates by Combining Multiple Genotyping Signals

Mohsen Katebi[1(✉)], Pedro Feijao[1], Julius Booth[1], Mehrdad Mansouri[1],
Sean La[1], Alex Sweeten[1], Reza Miraskarshahi[1], Matthew Nguyen[1],
Johnathan Wong[1], William Hsiao[2], Cedric Chauve[1], and Leonid Chindelevitch[1]

[1] Simon Fraser University,
8888 University Ave, Burnaby, BC V5A 1S6, Canada
mkatebi@sfu.ca
[2] British Columbia Centre for Disease Control,
655 West 12th Ave, Vancouver, BC V5Z 4R4, Canada

**Abstract.** In this paper we study the problem of clustering bacterial isolates into epidemiologically related groups from next-generation sequencing data. Existing methods for this problem mainly use a single genotyping signal, and either use a distance-based method with a pre-specified number of clusters, or a phylogenetic tree-based method with a pre-specified threshold. We propose PathOGiST, an algorithmic framework for clustering bacterial isolates by leveraging multiple genotypic signals and calibrated thresholds. PathOGiST uses different genotypic signals, clusters the isolates based on these individual signals with correlation clustering, and combines the clusterings based on the individual signals through consensus clustering. We implemented and tested PathOGiST on three different bacterial pathogens - *Escherichia coli*, *Yersinia pseudotuberculosis*, and *Mycobacterium tuberculosis* - and we conclude by discussing further avenues to explore.

**Keywords:** Bacterial pathogens · Whole-genome sequencing · Correlation clustering · Microbiology · Public health

## 1 Introduction

Partitioning the isolates of a bacterial pathogen into epidemiologically related groups is an important challenge in public health microbiology. Specifically, such a partitioning, which we will refer to as a *clustering*, can provide information on particularly transmissible strains (super-spreaders) and identify where an intervention such as active case finding may be particularly beneficial. In combination with additional metadata, such as geography or time of observation, such a clustering can also help identify rapidly growing groups (transmission hotspots),

© Springer Nature Switzerland AG 2020
C. Martín-Vide et al. (Eds.): AlCoB 2020, LNBI 12099, pp. 108–124, 2020.
https://doi.org/10.1007/978-3-030-42266-0_9

narrow down the potential origins of an outbreak (index case), and distinguish between recent and historical transmissions.

The clustering problem can leverage a variety of genotypic signals. Historically, fairly coarse genotypes such as VNTR (variable-number of tandem repeats, i.e. the number of copies of a set of pre-specified repeated regions in a strain) [29], PFGE (pulsed field gel electrophoresis) [15] and MLST (multi-locus sequence type, i.e. the alleles at a small number of pre-specified housekeeping genes) [18] have been the predominant mode of genotyping bacterial pathogens. These low-resolution signals, which we refer to as "fingerprints", could lead to incorrectly clustered strains [1] since unrelated bacterial isolates may happen to share identical fingerprints. With the advent of next-generation sequencing (NGS) [17], new genotypic signals have become available. These include SNP (single-nucleotide polymorphism) profiles [6], which can be identified at the whole-genome scale, and also wgMLST (whole-genome multi-locus sequence type) [19], which contains the alleles at all of the known genes in the organism of interest.

Methodologically, existing approaches fall into one of two categories. Some methods - including those inspired by and used in metagenomics [24] - use a pure distance-based approach, whereby a sequence similarity cutoff threshold is chosen, and any pair of sequences whose similarity exceeds it are considered to be in the same cluster, with a transitive closure operator applied to ensure the result is a valid partition. Alternatively, such methods may simply apply a standard clustering method, such as hierarchical clustering, to the pairwise distance matrix; in this case, the number of clusters is typically specified in advance [5]. Other methods - which tend to be more computationally expensive - leverage a phylogenetic tree reconstructed from the data to define clusters [2,11]. They also typically require a similarity threshold, but may be less sensitive to outlier isolates or to homoplasy, i.e. convergent evolution.

The majority of existing approaches for clustering bacterial isolates use a single genotypic signal, typically one of the higher-resolution ones, in isolation [12]. However, in this paper we argue for the principled combination of both low-resolution as well as high-resolution genotypic signals. The framework we propose here, called PathOGiST, innovates in several key ways. First, it leverages multiple genotypic signals extracted from NGS data. They can be further subdivided according to granularity into coarse and fine signals; the former get penalized only for grouping together isolates with different genotypes, not for splitting isolates with similar genotypes, while the latter get penalized for both of these. Second, it is based on a distance threshold, but does not apply a transitive closure operator to the similarity graph, or require a pre-specified number of clusters. Instead, it makes use of the *correlation clustering* paradigm, which tries to minimize the number of pairs of distant isolates within clusters while minimizing the number of pairs of close isolates between clusters. Third, it can be calibrated to different bacterial pathogens and genotyping signals, although we also provide an automatic threshold detector based on the distribution of pairwise distances between isolates.

Our results demonstrate that, when applied to a selection of three bacterial pathogens with annotated datasets publicly available - *Escherichia coli*, *Yersinia pseudo-tuberculosis*, and *Mycobacterium tuberculosis* - PathOGiST performs with a higher accuracy than recently published existing methods in most cases, both in terms of its ARI (adjusted Rand index) as well as CP (cluster purity). Our paper establishes that the use of calibrated thresholds and multiple genotypic signals can lead to an accurate clustering of bacterial isolates for public health epidemiology.

## 2    Methods

The goal of our approach is to cluster pathogen isolates from whole-genome sequencing data by using different genotyping approaches, alone and in combination. Each cluster should ideally represent a set of isolates related by an epidemiological transmission chain. We assume that we are given as input several matrices recording the pairwise distances between the isolates, one per genotyping signal. The algorithm proceeds in two stages. We first compute a clustering of the isolates for each distance matrix, and then compute a consensus of these separate clusterings. For the first step, we rely on *correlation clustering* [3], which we describe in Sect. 2.1. For the second step, we use a modified approach to the consensus clustering problem [4], also based on a correlation clustering formulation, described in Sect. 2.2. The whole process is illustrated in Fig. 1.

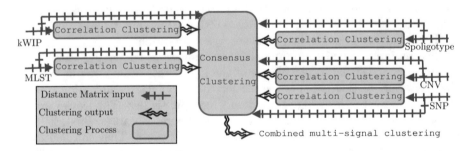

**Fig. 1.** PathOGiST starts by computing clusters based on single distance signals using correlation clustering. Then we run consensus clustering on the outputs of the correlation clustering.

### 2.1    Correlation Clustering

Let $G$ be an undirected complete weighted graph with vertices $V$ and edges $E$. Let $W : E \to \mathbb{R}$ be the edge-weighting function, which is positive for edges connecting vertices (representing isolates) that are similar and negative for those connecting dissimilar vertices. Correlation clustering aims to partition the vertices into disjoint clusters $C_1, C_2, \ldots, C_N$ where $N \leq n$. Let $I$ be the set of edges whose endpoints lie in the same cluster and let $J = E - I$ be the set of edges whose endpoints lie in different clusters. The goal of the *minimum correlation clustering problem* is to find a clustering that minimizes the total weight of the edges in $I$ with negative weight minus the total weight of the edges in $J$ with positive weight:

$$\underset{C_1, C_2, \ldots, C_N}{\arg \min} \sum_{\substack{e \in I \\ W(e) < 0}} W(e) - \sum_{\substack{e \in J \\ W(e) > 0}} W(e)$$

In this work, we perform the construction of the weighted graph $G$ from a distance matrix. Given a distance matrix $D$ on the input elements (graph vertices) such that $d_{ij}$ is the distance between elements $i$ and $j$, we define $s_{ij} = T - d_{ij}$, where $T$ is a *distance threshold*, intuitively meaning that if $d_{ij} < T$, $i$ and $j$ are considered similar, while $d_{ij} > T$ means that $i$ and $j$ are considered dissimilar. We use $s_{ij}$ as the weight of the edge between vertices $i$ and $j$ in $G$.

By defining binary variables $x_{ij}$ such that $x_{ij} = 0$ if $i$ and $j$ are in the same cluster and $x_{ij} = 1$ otherwise, we can write the minimum correlation clustering objective function as

$$f(x) = \sum_{s_{ij} > 0} s_{ij} x_{ij} - \sum_{s_{ij} < 0} s_{ij}(1 - x_{ij}) = \sum s_{ij} x_{ij} - \sum_{s_{ij} < 0} s_{ij}.$$

Since the second term is constant, the minimum correlation clustering problem can be solved optimally with the following Integer Linear Program (ILP):

$$\underset{x}{\text{minimize}} \quad \sum s_{ij} x_{ij} \qquad (1)$$

$$\text{s.t.} \quad x_{ik} \leq x_{ij} + x_{jk} \quad \text{for all } i, j, k$$

$$x_{ij} \in \{0, 1\} \quad \text{for all } i, j$$

Here, the inequality constraints (which we call the "triangle inequality" constraints) together with the binary constraints ensure that the assignment is transitive [3]. Indeed, if $x_{ij} = 0$ and $x_{jk} = 0$, they enforce that $x_{ik} = 0$.

## C4: A Fast Parallel Heuristic for the Correlation Clustering Problem.
Solving the ILP in Eq. (1) can be time consuming and can require a large amount of memory due to the quadratic number of variables and cubic number of constraints. For this reason, we additionally implemented the faster C4 algorithm, a parallel algorithm that guarantees a 3-approximation ratio of the optimal objective function of correlation clustering in the special case of metric distances (i.e. when the $s_{ij}$ satisfy the triangle inequality $s_{ik} \leq s_{ij} + s_{jk}$) [26].

Our results show that this algorithm is remarkably fast and quite accurate on the input graphs we tested. However, it is non-deterministic, as it depends on the initial permutation of the vertices. With this in mind, we run C4 multiple times with random initial permutations and compute the objective value for each solution. Then, among those solutions, we choose the one that minimizes the objective function. Our experiments show that in practice, this works very well and most of the time is able to find the optimum or near-optimum solution.

**Solving the Minimum Correlation Clustering Problem Exactly.** In order to solve the minimum correlation clustering problem exactly, while coping with the cubic number of linear constraints, we employed two approaches.

First, recalling that we often get a near-optimum solution from C4, we use it as a warm start to the problem by supplying it to the ILP solver.

Second, rather than creating the ILP with all the constraints right away, we iteratively add the constraints as follows. According to Eq. (1), for every index triple $(i, j, k)$ the ILP has 3 constraints on the decision variables $x_{ij}$, $x_{jk}$, $x_{ik}$:

$$x_{ik} \leq x_{ij} + x_{jk}, \ x_{ij} \leq x_{ik} + x_{jk}, \ x_{jk} \leq x_{ik} + x_{ij}.$$

To provide the intuition for our second heuristic, assume that all three similarities between elements $i, j, k$ are positive. This implies that the elements $i$, $j$, and $k$ are similar to each other, so are more likely to belong to the same cluster. In this case, the three variables $x_{ik}$, $x_{ij}$, and $x_{jk}$ will likely be assigned the value 0 and satisfy the inequalities. On the other hand, all three similarities being negative implies that elements $i$, $j$, and $k$ are likely to be in different clusters, which would set these variables to 1 and again satisfy the inequalities.

Taking this into account, we use an approach inspired by constraint generation [8], and start by only including constraints induced by element triples whose set of similarities contain both positive and negative edges and solve this trimmed-down ILP. We then check all the excluded constraints in the solution to see whether any of them is violated. If none is violated, then the current solution is also an optimum solution for the original ILP and we are done. Otherwise, we add all the constraints that are not satisfied by the current solution to the ILP and solve the modified ILP again. We repeat this process until no violated constraint remains.

In most experiments (225 out of 235 experiments), we observe that no violated constraints have been found. Almost all of the other cases only required one extra iteration to find a solution that satisfied all the constraints. The average number of iterations was 1.102.

## 2.2   Consensus Clustering

Given a set of clusterings and a measure of distance between clusterings, the *consensus clustering problem* aims to find a clustering minimizing the total distance to all input clusterings. A simple distance between two clusterings $\pi_1$ and $\pi_2$ is the number of elements clustered differently in $\pi_1$ and $\pi_2$, that is, the number of pairs of elements co-clustered in $\pi_1$ but not co-clustered in $\pi_2$, or vice versa.

Representing a clustering $x$ by a quadratic number of binary variables ($x_{ij} = 0$ if and only if $i, j$ are co-clustered), the distance between $x$ and a clustering $\pi$ is given by the formula

$$d(x, \pi) = \sum_{\pi_{ij}=1} w_{ij}(1 - x_{ij}) + \sum_{\pi_{ij}=0} w_{ij}x_{ij} = \sum_{i,j} s_{ij}x_{ij} + \sum_{x_{ij}=1} w_{ij}, \quad (2)$$

where a weight $w_{ij}$ is assigned to each pair of elements $i$ and $j$ penalizing any clustering decision in $x$ that differs from $\pi$, and we define $s_{ij} := (-1)^{\pi_{ij}} w_{ij}$.

Notice that solving the minimum consensus problem for a given set of clusterings $\pi^{(1)}, \ldots, \pi^{(n)}$ is equivalent to solving a minimum correlation clustering problem with the matrix $S$ defined as

$$s_{ij} = \sum_{\left\{k \mid \pi_{ij}^{(k)} = 0\right\}} w_{ij}^{(k)} - \sum_{\left\{k \mid \pi_{ij}^{(k)} = 1\right\}} w_{ij}^{(k)} = \sum_{k=1}^{n} (-1)^{\pi_{ij}^{(k)}} w_{ij}^{(k)} \tag{3}$$

**Consensus Clustering with Different Granularities.** An important feature of our problem is that the different genotyping signals we consider might not cluster the isolates with the same granularity. For example, it was shown in [22] that when clustering *Mycobacterium tuberculosis* isolates using SNPs, MLST, CNVs and spolygotypes, the latter two genotyping signals result in coarser clusters than the former two. For this reason, we assume that the input clusterings can be of different granularities. In this setting, we want to avoid penalizing the differences between a finer clustering $\pi$ and a coarser clustering $\pi'$, and we introduce the following asymmetric distance: $d(\pi, \pi') = |\pi - \pi'|$. In this case, assuming $\pi$ is the coarser clustering and $\pi'$ the finer one, we penalize only those pairs that are co-clustered in $\pi$ but not in $\pi'$.

Then, given the clusterings $\pi^{(1)}, \ldots, \pi^{(m)}$ and a subset $F$ of these clusterings, representing the clusterings with the finer resolution, the *finest consensus clustering* problem is to find a clustering $x$ that minimizes the total distance between $x$ and all input clusterings, where

$$d(x, \pi) = \begin{cases} \displaystyle\sum_{\pi_{ij}=1} w_{ij}(1 - x_{ij}) + \sum_{\pi_{ij}=0} w_{ij}x_{ij}, & \text{if } \pi \in F \\ \displaystyle\sum_{\pi_{ij}=0} w_{ij}x_{ij}, & \text{otherwise} \end{cases} \tag{4}$$

We can then reformulate this problem as a minimum correlation clustering again, with matrix $S$ defined by

$$s_{ij} = \sum_{\left\{k \mid \pi_{ij}^{(k)} = 0\right\}} w_{ij}^{(k)} - \sum_{\left\{k \mid \pi_{ij}^{(k)} = 1, \pi^{(k)} \in F\right\}} w_{ij}^{(k)} \tag{5}$$

**Selecting Appropriate Weights for Consensus Clustering.** There might be many meaningful ways of defining the weights $w_{ij}^{(k)}$ used in the previous equations. If we assume that a clustering $\pi$ was inferred based on a distance matrix $D$, normalized such that $0 \leq d_{ij} \leq 1$, we can define $w_{ij}$ as

$$w_{ij} = \begin{cases} d_{ij}, & \text{if } \pi_{ij} = 1 \\ 1 - d_{ij}, & \text{otherwise} \end{cases} \tag{6}$$

The reasoning behind this definition is that if $\pi_{ij} = 1$ ($i, j$ are not co-clustered in $\pi$), then the distance $d_{ij}$ should be large, therefore it is a good penalty for co-clustering $i, j$ in $x$. On the other hand, if $\pi_{ij} = 0$, $d_{ij}$ can be expected to be small, which means that $1 - d_{ij}$ is a better candidate for the penalty of choosing $x_{ij} = 1$. The distance between two clusterings (Eq. (2)) can then be written as

$$d(x, \pi) = \sum_{\pi_{ij}=1} d_{ij}(1 - x_{ij}) + \sum_{\pi_{ij}=0} (1 - d_{ij})x_{ij} \tag{7}$$

and Eq. (3) becomes

$$s_{ij} = \sum_{\{k|\pi_{ij}^{(k)}=0\}} \left(1 - d_{ij}^{(k)}\right) - \sum_{\{k|\pi_{ij}^{(k)}=1\}} d_{ij}^{(k)} = \Pi_{ij} - D_{ij} \tag{8}$$

where $\Pi_{ij} = \left|\left\{k|\pi_{ij}^{(k)} = 0\right\}\right|$ and $D = \sum_{k=1}^{n} d_{ij}^{(k)}$.

We can naturally combine the weighting with the different granularities within a single formulation. In summary, the finest consensus clustering problem with weights can be formulated as a minimum correlation clustering problem, and thus solved by the algorithms described in Sect. 2.1.

## 2.3   Evaluation

To evaluate our methods for clustering, we compute two measures between our clustering and a ground truth clustering: Adjusted Rand Index (ARI) and Cluster Purity (CP).

The adjusted Rand index is a measure that computes how similar the clusters are to the ground truth. It is the corrected-for-chance version of the Rand index which is the percentage of correctly clustered elements. It can be computed using the following formula:

$$ARI = \frac{\sum_{ij} \binom{n_{ij}}{2} - \left[\sum_i \binom{a_i}{2} \sum_j \binom{b_j}{2}\right] / \binom{n}{2}}{\frac{1}{2}\left[\sum_i \binom{a_i}{2} + \sum_j \binom{b_j}{2}\right] - \left[\sum_i \binom{a_i}{2} \sum_j \binom{b_j}{2}\right] / \binom{n}{2}}$$

where $n_{ij}$, $a_i$, $b_j$ are values, row sums, and column sums from the contingency table [13].

Cluster Purity is another measure of similarity between two data clusterings. To compute, assign each cluster to the most common ground truth cluster in it. Then, count the number of correctly assigned data points and divide by the total number of data points. Formally:

$$CP(C, G) = \frac{1}{N} \sum_k \max_j |c_k \cap g_j|$$

where $N$ is the number of data points, $C = \{c_1, c_2, \ldots, c_K\}$ is the set of clusters and $G = \{g_1, g_2, \ldots, g_J\}$ is the set of ground truth clusters [20].

# 3   Results

## 3.1   Datasets and Genotyping Methods

We used three published datasets for three pathogens, *Escherichia coli* [14], *Mycobacterium tuberculosis* [10], *Yersinia pseudotuberculosis* [30], and a simulated dataset taken from [16]. Several genotyping signals were extracted from the WGS data: multilocus sequence typing (MLST) using MentaLiST pipeline [7], single nucleotide polymorphisms (SNP) using Snippy [28], copy number variants (CNV) using Prince [21], *k*-mer weighted inner products (kWIP) using kWIP [23], and spacer oligonucleotide typing (Spoligotyping) using SpoTyping [31] (Table 1).

**Table 1.** Datasets and genotyping summary

| Dataset | Number of isolates | Genotyping signals | | | | |
|---|---|---|---|---|---|---|
| | | SNP | MLST | kWIP | CNV | SpoTyping |
| *E. coli* | 1509 | ✓ | ✓ | ✓ | ✗ | ✗ |
| *M. tuberculosis* (MTB) | 1377 | ✓ | ✓ | ✓ | ✓ | ✓ |
| *Y. pseudotuberculosis* (Yp) | 163 | ✓ | ✓ | ✓ | ✗ | ✗ |
| *Simulated Data* (SD) | 96 | ✓ | ✓ | ✓ | ✗ | ✗ |

For each genotyping signal, in order to apply our correlation clustering algorithm, we needed to determine a threshold $T$ to decide which pairs of isolates should be considered similar. To do so, we consider the pairwise distance distribution for each signal, choosing a threshold range that covers the first valley in the distribution, under the assumption that the first peak likely indicates distances between isolates belonging to the same cluster. The resulting threshold ranges and steps are described in Appendix (Table 4).

In our experiments, for each sample, each signal and each threshold, we ran our two algorithms for solving the minimum correlation clustering problem, the C4 approximation algorithm (with multiple runs) and the exact ILP using delayed constraint generation. Then we ran the consensus clustering algorithm, again using both methods.

## 3.2   Single Signal Genotyping

The *E. coli* dataset contains 1509 isolates collected from across England and spans an 11-year period. The distance distribution for each genotyping signal (MLST, SNP, kWIP) is shown in the Appendix (Fig. 2).

The second dataset contains 163 isolates of *Y. pseudotuberculosis* mostly collected from New Zealand [30]. We applied the same genotyping methods as for the *E. coli* dataset to this one. The results are presented in Appendix (Fig. 3). For *E. coli* and *Y. pseudotuberculosis*, we consider the MLST groups determined in their respective studies [14,30] as the ground truth, and use them to calculate the ARI and CP values. For the simulated data, since the authors suggest [16] BAPS for building the true phylogenetic tree and clustering, we used the RhierBAPS results as the ground truth for this dataset instead of MLST.

The isolates of the *M. tuberculosis* dataset were obtained from pediatric patients in British Columbia, Canada, and were collected between 2005 and 2014. We used a subset of 1377 isolates, all of which underwent WGS. In addition to SNP, MLST and kWIP information, we considered two additional genotyping signals, CNV and Spolygotyping. For *M. tuberculosis*, due to the lack of MLST groups, we use the strain's lineage, a proxy for its geographic origin [27]. For this dataset, the ARI would not be informative since a lineage is a coarse grouping largely uninformative of the underlying epidemiology, and should be split into multiple clusters. Thus, we only calculate and report the CP in this case. The results for this dataset are illustrated in Appendix (Fig. 4).

We can observe that for all the pathogens and genotyping signals we consider, there is a relatively clear threshold that falls within the chosen range which results in high accuracy clusters, with ARI and CP values above 0.8, and often close to 1. The only exceptions concern the *M. tuberculosis* dataset with SNPs, MLST and kWIP, although the CP statistics is notoriously less robust than the ARI. Moreover, most of the time, around the best thresholds, the clustering obtained with the exact ILP method results in accuracy measures that are either very close to those of the C4 method or slightly better.

### 3.3   Comparison of the C4 and Exact ILP Methods

The ILP generally gives more accurate results and is a deterministic method. However, its running time and memory usage depend a lot on the size of the dataset. For example, while it is able to cluster the smaller *Y. pseudotuberculosis* dataset in less than a minute, it takes more than three hours to find clusters for larger datasets at some threshold values. On the other hand, the C4 heusritic is significantly faster and requires much less memory even on the larger datasets, as shown in Table 2. However, it is not deterministic, and random restarts may give slightly different, incompatible results. To evaluate the C4 heuristic performance, we compared the objective values of solutions found by C4 and by the exact ILP. In most cases, C4 performs very well and finds a solution whose objective value is close to the optimal (Table 2).

Furthermore, the objective value of CPLEX is affected by the tolerance parameter; when the gap between the lower and upper bound is less than a certain fraction $\epsilon$, set to $10^{-6}$ by default, the optimization is stopped. In this case, we see that because the magnitude of the objective function is fairly large, it is possible for the C4 method to obtain a better objective function than CPLEX.

**Table 2.** Average running time (in seconds) and memory footprint (in gigabytes); ILP and C4 objective value comparison.

| Dataset | Time (s) | | Memory (GB) | | Objective value | | |
|---|---|---|---|---|---|---|---|
| | C4 | ILP | C4 | ILP | ILP | C4: mean | C4: std |
| *E. coli* | 698 | 7282 | 0.22 | 193.72 | $-1.9068 \times 10^{10}$ | $-1.9074 \times 10^{10}$ | $3.8817 \times 10^{5}$ |
| *M. tuberculosis* | 572 | 10437 | 0.20 | 298.87 | $-2.9445 \times 10^{8}$ | $-2.8844 \times 10^{8}$ | $2.4238 \times 10^{3}$ |
| *Y. pseudotuberculosis* | 14 | 15 | 0.13 | 0.81 | $-4.1601 \times 10^{7}$ | $-4.1594 \times 10^{7}$ | $1.3238 \times 10^{3}$ |

## 3.4   Comparison with Existing Clustering Methods

The results from PathOGiST were compared to those generated by two recent methods developed for clustering WGS datasets, Phydelity [11] and TreeCluster [2]; both of them are based on phylogenetic trees. To infer a phylogeny for our datasets, we first calculated a pair-wise distance matrix using Mash [25], then we ran the popular and widely used BIONJ [9] variant of the neighbor joining algorithm on the distance matrix. After we inferred phylogenetic trees, we ran Phydelity and TreeCluster with their default settings. In order to pick a single threshold for each genotyping signal-pathogen combination in PathOGiST, we chose the threshold resulting in the best ARI (CP for *M. tuberculosis*) among all the options. These thresholds are set as the default thresholds for these genotyping signal-pathogen combinations, but can be overridden by the user. Table 5 (Appendix) shows the chosen optimal threshold for each dataset and genotyping signal.

**Table 3.** ARI (Adjusted Rand Index) and CP (Cluster Purity) computed for different methods and genotyping signals

| Method | *E. coli* | | *Y. pseudotuberculosis* | | *M. tuberculosis* | | *Simulated Data* | |
|---|---|---|---|---|---|---|---|---|
| | ARI | CP | ARI | CP | ARI | CP | ARI | CP |
| Phydelity | 0.76 | 0.93 | 0.23 | 0.94 | - | 0.92 | 0.238 | 0.645 |
| TreeCluster | 0.08 | 0.96 | 0.01 | 0.90 | - | 0.74 | 0.940 | 0.562 |
| **PathOGiST** | | | | | | | | |
| ILP: SNP | **0.92** | **1.0** | **0.96** | **0.98** | - | 0.56 | 0.970 | 0.979 |
| ILP: MLST | 0.90 | 0.95 | 0.94 | 0.94 | - | 0.95 | 0.969 | 0.968 |
| ILP: kWIP | 0.90 | **1.0** | **0.96** | 0.94 | - | 0.57 | **0.973** | **0.989** |
| ILP: CNV | - | - | - | - | - | **1.0** | - | - |
| ILP: SpoTyping | - | - | - | - | - | 0.92 | - | - |
| ILP: Consensus | 0.91 | 0.85 | **0.96** | 0.97 | - | 0.57 | **0.973** | **0.989** |
| C4: SNP | **0.92** | **1.0** | **0.96** | **0.98** | - | 0.57 | **0.973** | **0.989** |
| C4: MLST | 0.90 | 0.95 | 0.94 | 0.94 | - | 0.95 | 0.969 | 0.968 |
| C4: kWIP | 0.90 | 0.99 | **0.96** | 0.94 | - | 0.60 | **0.973** | **0.989** |
| C4: CNV | - | - | - | - | - | **1.0** | - | - |
| C4: SpoTyping | - | - | - | - | - | 0.92 | - | - |
| C4: Consensus | 0.91 | 0.86 | **0.96** | 0.97 | - | 0.47 | **0.973** | **0.989** |

Having clustering outputs of the single signal correlation clustering algorithm with chosen default thresholds, we ran consensus clustering for each pathogen with all their available genotyping signals. We considered SNP clustering as the finest because it provides a higher resolution signal comparing to other genotyping signals. The results are described in Table 3. The main observation is that in all cases, but *M. tuberculosis*, the consensus clustering ARI is close to the best ARI obtained by a single genotyping signal, showing that our approach indeed removes the need to chose a single signal for clustering.

## 4    Conclusion

In this paper we described PathOGiST, an algorithmic framework for clustering bacterial isolates. One of our key contributions is to introduce the paradigms of correlation clustering and consensus clustering for the analysis of bacterial pathogens, together with two implementations - one exact and one heuristic - of correlation clustering algorithms, tailored to the problem at hand. Our experimental results suggest that our approach allows to compute a very accurate, often close to optimal, clustering without having to determine an optimal genotyping signal.

In the future, we hope to address several challenges. The first issue is the risk of overfitting, as the calibration of the threshold relies on the correlation clustering results' comparison to a gold standard. However, we also provide an automatic threshold detector. Our results demonstrate that our approach has the potential to provide reliable clusters.

Second, instead of a single output, a multi-scale or hierarchical representation of the clusters may be helpful in order to provide the user with the flexibility of deciding on their own clustering granularity. Moreover, some metadata, such as collection time or geographic location, may be fruitfully incorporated into the clustering approach in order to better inform the resolution of some groups of isolates.

Finally, due to the lack of existing tools for simulating multiple genotyping signals, we considered MLST as our gold standard. This may not represent the correct clustering, but is the best available among the individual genotyping signals.

Despite these challenges, we believe that PathOGiST is a first step in the right direction, and we hope that it will generate an impetus to further explore the problem of clustering bacterial isolates.

# A    Appendix Tables

**Table 4.** Ranges and steps for threshold values. For each experiment we ran the PathOGiST with different thresholds iteratively. We increased threshold by the step starting from beginning of the range through its end.

| Dataset | SNP | | MLST | | kWIP | | CNV | | SpoTyping | |
|---------|-----|------|------|------|------|------|------|------|------|------|
| | Range | Step | Range | Step | Range | Step | Range | Step | Range | Step |
| *E. coli* | $(0, 43000]$ | 2150 | $(0, 600]$ | 20 | $[0.21, 0.75]$ | 0.03 | - | - | - | - |
| MTB | $(0, 500]$ | 25 | $(0, 500]$ | 25 | $[0.125, 0.5]$ | 0.025 | $(0, 50]$ | 2.5 | $(0, 13]$ | 0.65 |
| Yp | $(0, 40000]$ | 2000 | $(0, 600]$ | 20 | $[0.175, 0.7]$ | 0.025 | - | - | - | - |
| SD | $(0, 8000]$ | 400 | $(0, 400]$ | 20 | $[0.26, 0.4]$ | 0.02 | - | - | - | - |

**Table 5.** Best clustering thresholds per dataset and genotyping signal.

| Dataset | SNP | MLST | kWIP | CNV | SpoTyping |
|---------|-----|------|------|-----|-----------|
| *E. coli* | 17200 | 400 | 0.66 | - | - |
| *M. tuberculosis* | 500 | 475 | 0.5 | 50 | 13 |
| *Y. pseudotuberculosis* | 6000 | 340 | 0.625 | - | - |
| *Simulated Data* | 1600 | 220 | 0.4 | - | - |

# B     Appendix Figures

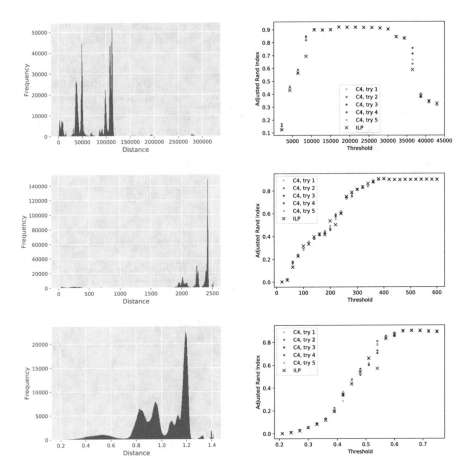

**Fig. 2.** Distance histograms and ARI results for *E. coli*. From top to bottom: SNP, MLST, kWIP.

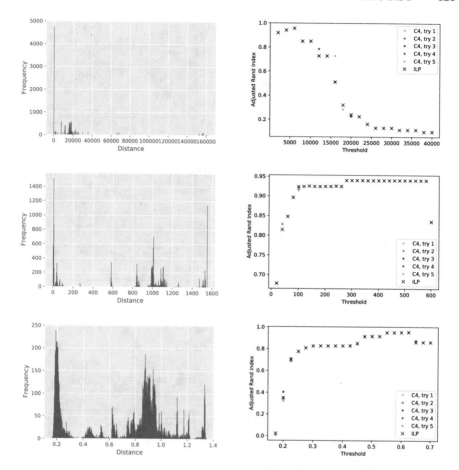

**Fig. 3.** Distance histograms and ARI results for the *Y. pseudotuberculosis* dataset. From top to bottom: SNP, MLST, kWIP.

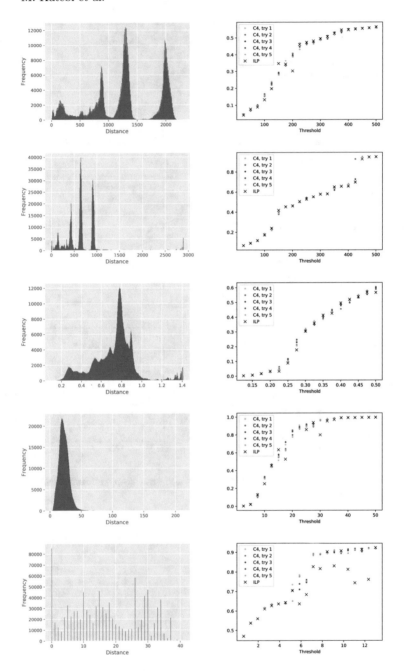

**Fig. 4.** Distance histograms and CP results for *M. tuberculosis*. From top to bottom: SNP, MLST, kWIP, CNV, Spolygotyping.

# References

1. Alaridah, N., Hallbäck, E.T., Tångrot, J., et al.: Transmission dynamics study of tuberculosis isolates with whole genome sequencing in southern Sweden. Sci. Rep. **9**(1), 4931 (2019)
2. Balaban, M., Moshiri, N., Mai, U., et al.: TreeCluster: clustering biological sequences using phylogenetic trees. bioRxiv (2019). https://doi.org/10.1101/591388
3. Bansal, N., Blum, A., Chawla, S.: Correlation clustering. Mach. Learn. **56**, 89–113 (2004)
4. Bonizzoni, P., Vedova, G.D., Dondi, R., Jiang, T.: On the approximation of correlation clustering and consensus clustering. J. Comput. Syst. Sci. **74**, 671–696 (2008)
5. Cheng, L., Connor, T.R., Sirén, J., et al.: Hierarchical and spatially explicit clustering of DNA sequences with BAPS software. Mol. Biol. Evol. **30**, 1224–1228 (2013)
6. Faison, W.J., et al.: Whole genome single-nucleotide variation profile-based phylogenetic tree building methods for analysis of viral, bacterial and human genomes. Genomics **104**(1), 1–7 (2014)
7. Feijao, P., Yao, H.T., Fornika, D., et al.: MentaLiST-a fast MLST caller for large MLST schemes. Microb. Genom. **4** (2018)
8. Dantzig, G., Fulkerson, R., Johnson, S.: Solution of a large-scale traveling salesman problem. Oper. Res. **2**, 393–410 (1954)
9. Gascuel, O.: BIONJ: an improved version of the NJ algorithm based on a simple model of sequence data. Mol. Biol. Evol. **14**(7), 685–695 (1997)
10. Guthrie, J.L., Delli Pizzi, A., Roth, D., et al.: Genotyping and whole-genome sequencing to identify tuberculosis transmission to pediatric patients in British Columbia, Canada, 2005–2014. J. Infect. Dis. **40**, 1–9 (2018)
11. Han, A.X., Parker, E., Maurer-Stroh, S., et al.: Inferring putative transmission clusters with Phydelity. bioRxiv (2019). https://doi.org/10.1101/477653
12. Hanage, W.P., Fraser, C., Spratt, B.G.: Sequences, sequence clusters and bacterial species. Philos. Trans. R. Soc. B: Biol. Sci. **361**(1475), 1917–1927 (2006)
13. Hubert, L., Arabie, P.: Comparing partitions. J. Classif. **2**, 193–218 (1985)
14. Kallonen, T., Brodrick, H.J., Harris, S.R., et al.: Systematic longitudinal survey of invasive Escherichia coli in England demonstrates a stable population structure only transiently disturbed by the emergence of ST131. Genome Res. **27**, 1437–1449 (2017)
15. Kaufmann, M.E.: Pulsed-field gel electrophoresis. In: Woodford, N., Johnson, A.P. (eds.) Molecular Bacteriology, pp. 33–50. Springer, Heidelberg (1998). https://doi.org/10.1385/0-89603-498-4:33
16. Lees, J.A., Kendall, M., Parkhill, J., Colijn, C., Bentley, S.D., Harris, S.R.: Evaluation of phylogenetic reconstruction methods using bacterial whole genomes: a simulation based study. Wellcome Open Res. **3** (2018)
17. Loman, N.J., Pallen, M.J.: Twenty years of bacterial genome sequencing. Nat. Rev. Microbiol. **13**(12), 787 (2015)
18. Maiden, M.C., Bygraves, J.A., Feil, E., et al.: Multilocus sequence typing: a portable approach to the identification of clones within populations of pathogenic microorganisms. PNAS **95**(6), 3140–3145 (1998)
19. Maiden, M.C., Van Rensburg, M.J.J., Bray, J.E., et al.: MLST revisited: the gene-by-gene approach to bacterial genomics. Nat. Rev. Microbiol. **11**(10), 728 (2013)

20. Manning, C., Raghavan, P., Schütze, H.: Introduction to information retrieval. Nat. Lang. Eng. **16**(1), 100–103 (2010)
21. Mansouri, M., Booth, J., Vityaz, M., et al.: PRINCE: accurate approximation of the copy number of tandem repeats. In: WABI 2018, pp. 20:1–20:13 (2018)
22. Meehan, C.J., Moris, P., Kohl, T.A., et al.: The relationship between transmission time and clustering methods in Mycobacterium tuberculosis epidemiology. EBioMedicine **37**, 410–416 (2018)
23. Murray, K.D., Webers, C., Ong, C.S., et al.: kWIP: the k-mer weighted inner product, a de novo estimator of genetic similarity. PLoS Comput. Biol. **13**, 1–17 (2017)
24. Nguyen, N.P., Warnow, T., Pop, M., White, B.: A perspective on 16S rRNA operational taxonomic unit clustering using sequence similarity. NPJ Biofilms Microbi. **2**, 16004 (2016)
25. Ondov, B.D., Treangen, T.J., Melsted, P., et al.: Mash: fast genome and metagenome distance estimation using minhash. Genome Biol. **17**(1), 132 (2016)
26. Pan, X., Papailiopoulos, D.S., Oymak, S., et al.: Parallel correlation clustering on big graphs. In: NIPS 2015, pp. 82–90 (2015)
27. Reed, M., Pichler, V., McIntosh, F., et al.: Major Mycobacterium tuberculosis lineages associate with patient country of origin. J. Clin. Microbiol. **47**, 1119–1128 (2009)
28. Seemann, T.: Snippy (2015). https://github.com/tseemann/snippy
29. Vergnaud, G., Pourcel, C.: Multiple locus variable number of tandem repeats analysis. In: Caugant, D. (ed.) Molecular Epidemiology of Microorganisms, pp. 141–158. Springer, Heidelberg (2009). https://doi.org/10.1007/978-1-60327-999-4_12
30. Williamson, D.A., Baines, S.L., Carter, G.P., et al.: Genomic insights into a sustained national outbreak of Yersinia pseudotuberculosis. Genome Biol. Evol. **8**, 3806–3814 (2017)
31. Xia, E., Teo, Y.Y., Ong, R.T.H.: SpoTyping: fast and accurate in silico mycobacterium spoligotyping from sequence reads. Genome Med. **8**(1), 19 (2016)

# TreeSolve: Rapid Error-Correction
# of Microbial Gene Trees

Misagh Kordi and Mukul S. Bansal[(⊠)]

Department of Computer Science and Engineering,
University of Connecticut, Storrs, USA
{misagh.kordi,mukul.bansal}@uconn.edu

**Abstract.** Gene tree reconstruction is an important problem in phylogenetics. However, gene sequences often lack sufficient information to confidently distinguish between competing gene tree topologies. To overcome this limitation, the best gene tree reconstruction methods use a known species tree topology to guide the reconstruction of the gene tree. While such *species-tree-aware* gene tree reconstruction methods have been repeatedly shown to result in vastly more accurate gene trees, the most accurate of these methods often have prohibitively high computational costs.

In this work, we introduce a highly computationally efficient and robust species-tree-aware method, named *TreeSolve*, for microbial gene tree reconstruction. TreeSolve works by collapsing weakly supported edges of the input gene tree, resulting in a non-binary gene tree, and then using new algorithms and techniques to optimally resolve the non-binary gene trees with respect to the given species tree in an appropriately and dynamically constrained search space. Using thousands of real and simulated gene trees, we demonstrate that TreeSolve significantly outperforms the best existing species-tree-aware methods for microbes in terms of accuracy, speed, or both. Crucially, TreeSolve also implicitly keeps track of multiple optimal gene tree reconstructions and can compute either a single best estimate of the gene tree or multiple distinct estimates. As we demonstrate, aggregating over multiple gene tree candidates helps distinguish between correct and incorrect parts of an error-corrected gene tree. Thus, TreeSolve not only enables rapid gene tree error-correction for large gene trees without compromising on accuracy, but also enables accounting of inference uncertainty.

**Keywords:** Phylogenetics · Microbial evolution · Gene tree reconstruction · Horizontal gene transfer

## 1 Introduction

One of the most fundamental tasks in studying gene family evolution is the construction of a *gene tree* showing the evolutionary relationships among individual genes from that gene family. However, it is well known that gene trees can be

© Springer Nature Switzerland AG 2020
C. Martín-Vide et al. (Eds.): AlCoB 2020, LNBI 12099, pp. 125–139, 2020.
https://doi.org/10.1007/978-3-030-42266-0_10

very hard to reconstruct accurately and there is often considerable uncertainty in gene tree topologies reconstructed using gene sequences alone [2,8–10]. To address the problem of gene tree error, many *species-tree-aware* methods have been developed for reconstructing or error-correcting gene trees. These methods make use of a known species tree and a phylogenetic reconciliation model that makes it possible to extract topological information from the species tree and use it to guide gene tree inference. In this work, we focus specifically on the reconstruction of microbial gene trees, where the relevant phylogenetic reconciliation model is the Duplication-Transfer-Loss (DTL) reconciliation which models the evolution of gene trees within species trees through speciation, gene duplication, gene loss, and horizontal gene transfer. Given its importance to understanding microbial evolution, the DTL reconciliation problem has been widely studied, e.g., [1,3,4,7,11–16], and all existing species-tree-aware methods for microbial gene trees are based on DTL reconciliation or its variants. Existing species-tree-aware methods for microbial gene trees include AnGST [3], MowgliNNI [9], ALE [15], PrIME-DLTRS [12], TreeFix-DTL [2], TERA [11], and ecceTERA [5]. Amongst all these methods, TreeFix-DTL [2] and ecceTERA [5] have been shown to be among the most accurate. Both TreeFix-DTL and ecceTERA are *gene tree error-correction* methods that take as input a previously reconstructed sequence-only gene tree and error-correct it based on a given species tree. Note that ecceTERA also implements the amalgamation-based algorithm implemented in TERA [11]; however, in this manuscript, ecceTERA refers only to the implementation of the gene tree resolution algorithm from [5].

In this work, we introduce a new species-tree-aware method, *TreeSolve* (portmanteau of *Tree* and *Resolve*), for error-correction of microbial gene trees that significantly outperforms the best existing methods in terms of accuracy, speed, or both. TreeSolve builds upon two key ideas already used for microbial gene tree error-correction and combines and extends them in novel ways. The first of these two keys ideas is to handle gene tree uncertainty by collapsing all weakly supported edges in the input sequence-based gene tree, resulting in a non-binary gene tree, and then optimally resolving this non-binary gene trees by reconciling to the given species tree, e.g., [5,7,17]. The second key idea is the consideration of gene tree bootstraps or other replicates to constrain the search space for the final gene tree to only a biologically meaningful subset of the full search space [3,11,15]. While both of these ideas have been separately used before, TreeSolve combines and extends them to achieve improved speed and accuracy. Specifically, TreeSolve collapses weakly supported edges of the input gene tree, resulting in a non-binary gene tree, and then uses new algorithms and techniques to optimally resolve the non-binary gene trees with respect to the given species tree in a constrained search space defined by a collection of bootstrap/replicate gene tree. An important novel aspect of our algorithm is that it is *self-adaptive* in that it can automatically increase or decrease the search space by considering only those clades that appear in at least a certain fraction of the bootstrap/replicate gene trees (by default, the considered clades should appear in at least one of the bootstrap/replicate gene trees). This self-adaptability is required because, even with the constraints imposed by the gene tree boot-straps/replicates, the number of optimal resolutions can grow exponentially in the degree and number of non-binary nodes in the given non-binary gene tree. By dynamically increasing or decreasing the minimum support value required for the clades considered, the algorithm is guaranteed to be very efficient even on very large and highly non-binary gene trees while still maintaining its accuracy. Another key strength of TreeSolve is that it implicitly keeps track of multiple, equally optimal, gene tree resolutions; it can either output a single best estimate of the gene tree or it can output multiple distinct gene tree candidates ordered by their average bootstrap/replicate support values.

We compared the accuracy and runtime of TreeSolve against the two most accurate gene tree error-correction methods for microbial gene trees, TreeFix-DTL [2] and ecceTERA [5], using an extensive experimental study with thousands of real and simulated gene trees. TreeFix-DTL has been previously demonstrated to have greater accuracy than AnGST and MowgliNNI [2], and ecceTERA demonstrated to have either greater or comparable accuracy to ALE, TERA, and PrIME-DLTRS [5,11]). Furthermore, ecceTERA is among the fastest species-tree-aware methods currently available for microbial gene trees, and it is also the method conceptually most similar to TreeSolve. Our results demonstrate that (i) TreeSolve is orders of magnitude faster and far more scalable than TreeFix-DTL, while matching or exceeding it in accuracy on larger gene trees, (ii) TreeSolve is significantly more accurate than ecceTERA and has comparable running times, (iii) the self-adaptive algorithm implemented in TreeSolve is highly scalable and efficient and can be easily applied to large genome-scale datasets and gene trees having many hundreds of leaves, and (iv) aggregating over multiple gene tree candidates output by TreeSolve helps distinguish between correct and incorrect branches of an error-corrected gene tree. An implementation of TreeSolve is available from https://compbio.engr.uconn.edu/software/TreeSolve/.

## 2   Definitions and Preliminaries

We follow basic definitions and notation from [1] and [7]. Given a tree $T$, we denote its node, edge, and leaf sets by $V(T)$, $E(T)$, and $Le(T)$ respectively.

If $T$ is rooted, the root node of $T$ is denoted by $rt(T)$, the parent of a node $v \in V(T)$ by $pa_T(v)$, its set of children by $Ch_T(v)$, and the (maximal) subtree of $T$ rooted at $v$ by $T(v)$. The set of *internal nodes* of $T$, denoted $I(T)$, is defined to be $V(T) \setminus Le(T)$. For a rooted tree $T$, we define $\leq_T$ to be the partial order on $V(T)$ where $x \leq_T y$ if $y$ is a node on the path between $rt(T)$ and $x$. The partial order $\geq_T$ is defined analogously, i.e., $x \geq_T y$ if $x$ is a node on the path between $rt(T)$ and $y$. We say that $y$ is an *ancestor* of $x$, or that $x$ is a *descendant* of $y$, if $x \leq_T y$ (note that, under this definition, every node is a descendant as well as ancestor of itself). We say that $x$ and $y$ are *incomparable* if neither $x \leq_T y$ nor $y \leq_T x$. Given a non-empty subset $L \subseteq Le(T)$, we denote by $lca_T(L)$ the last common ancestor (LCA) of all the leaves in $L$ in tree $T$.

A rooted tree is *binary* if all of its internal nodes have exactly two children, and *non-binary* otherwise. An *internal edge* is an edge whose end points are both internal nodes in the tree. An internal edge $(x, pa_T(x))$ in tree $T$ can be *contracted* by removing $(x, pa_T(x))$ and creating new edges joining $pa_T(x)$ with $Ch_T(x)$, thereby yielding a new tree distinct from $T$. We say that a tree $T'$ is a *binary resolution* of $T$ if $T'$ is binary and $T$ can be obtained from $T'$ by contracting some (zero or more) internal edges. We denote by $\mathcal{BR}(T)$ the set of all binary resolutions of a rooted non-binary tree $T$. Given any node $x$ from $T$, we define the *out-degree* of $x$ to be the number of children of $x$.

For a rooted tree $T$ each node $v \in V(T)$, the *clade* $C_T(v)$ is defined to be the set of all leaf nodes in $T(v)$; i.e. $C_T(v) = Le(T(v))$. We denote the set of all clades of a rooted tree $T$ by $Clade(T)$. This concept can be extended to unrooted trees as follows. Suppose $T$ is an unrooted tree. Each edge $(u, v) \in E(T)$ defines a partition of the leaf set of $T$ into two disjoint subsets $Le(T_u)$ and $Le(T_v)$, where $T_u$ is the subtree containing node $u$ and $T_v$ is the subtree containing node $v$, obtained when edge $(u, v)$ is removed from $T$. We call $Le(T_u)$ and $Le(T_v)$ the *clusters* of $T$ induced by edge $(u, v)$, and denote the set of all clusters in an unrooted tree $T$ by $Cluster(T)$.

In this work, we will consider both rooted and unrooted trees. However, unless otherwise specified, the term *tree* refers to a rooted tree.

A *species tree* is a tree that depicts the evolutionary relationships of a set of species. Given a gene family from a set of species, a *gene tree* is a tree that depicts the evolutionary relationships among the sequences encoding only that gene family in the given set of species. Gene trees may be either binary or non-binary while the species tree is always assumed to be binary. Throughout this work, we denote the gene tree and species tree under consideration by $G$ and $S$, respectively. If $G$ is restricted to be binary we refer to it as $G^B$ and as $G^N$ if it is restricted to be non-binary. We assume that each leaf of the gene tree is labeled with the species from which that gene was sampled. This labeling defines a *leaf-mapping* $\mathcal{L}_{G,S} \colon Le(G) \to Le(S)$ that maps a leaf node $g \in Le(G)$ to that unique leaf node $s \in Le(S)$ that has the same label as $g$. Note that gene trees may have more than one gene sampled from the same species.

**Reconciliation and DTL-scenarios.** A binary gene tree can be reconciled with a species tree by mapping the gene tree into the species tree. A Duplication-Transfer-Loss scenario (DTL-scenario) [1,16] for $G^B$ and $S$ characterizes the mappings of $G^B$ into $S$ that constitute a biologically valid reconciliation. Essentially, DTL-scenarios map each gene tree node to a unique species tree node and designate each gene tree node as representing either a speciation, duplication, or transfer event. A formal definition of DTL-scenario appears in [1]. DTL-scenarios correspond naturally to reconciliations and it is straightforward to infer the reconciliation of $G^B$ and $S$ implied by any DTL-scenario. Given a DTL-scenario, one can directly count the number of duplications, transfers, and losses invoked by the corresponding reconciliation [1].

Let $P_\Delta$, $P_\Theta$, and $P_{loss}$ denote the non-negative costs associated with duplication, transfer, and loss events, respectively. The *reconciliation cost* of a DTL-scenario is defined to be the total cost of all duplication, transfer, and loss events invoked by that DTL-scenario. A most parsimonious reconciliation is one that has minimum reconciliation cost.

**Definition 1 (MPR).**  *Given $G^B$ and $S$, along with $P_\Delta$, $P_\Theta$, and $P_{loss}$, a most parsimonious reconciliation (MPR) for $G^B$ and $S$ is a DTL-scenario with minimum reconciliation cost.*

**Optimal Gene Tree Resolution.** TreeSolve works by first converting the given binary gene tree into a non-binary gene tree by collapsing weakly supported edges (based on a user-provided threshold), and then optimally resolving this non-binary gene tree based on the species tree under appropriate topological constraints. A closely related problem formulation that has been previously studied is that of optimal gene tree resolution (OGTR) under DTL reconciliation [6,7]. In the OGTR problem, given non-binary gene tree $G^N$ and a species tree, one must find a binary resolution $G^B$ of $G^N$ such that an MPR of $G^B$ with $S$ has smallest reconciliation cost. Moreover, since there may be more than one optimal binary resolution of $G^N$, the desired formulation of the problem is to find *all* optimal resolutions of $G^N$. This leads to the following computational problem [7].

**Problem 1 (OGTR-All).**  *Given $G^N$ and $S$, along with $P_\Delta$, $P_\Theta$, and $P_{loss}$, the All Optimal Gene Tree Resolutions (OGTR-All) problem is to compute the set $\mathcal{OR}(G^N)$ of all optimal binary resolutions of $G^N$ such that, for any $G^B \in \mathcal{OR}(G^N)$, an MPR of $G^B$ and $S$ has the smallest reconciliation cost among all gene trees in $\mathcal{BR}(G^N)$.*

The OGTR-All problem is known to be NP-hard [6] (even for computing a single optimal resolution), and existing algorithms are limited to solving instances in which the maximum out-degree in $G^N$ is small [7].

**Constrained Optimal Gene Tree Resolution.** In addition to its very high computational time complexity, which greatly limits its applicability, the OGTR-All problem ignores sequence information and is therefore prone to over-fitting the gene tree to the species tree. TreeSolve addresses both these limitations by constraining the set of binary resolutions of $G^N$ that can be considered. Specifically, TreeSolve allows all binary resolutions that are supported by the sequence data and disallows those that are unsupported. To achieve this goal TreeSolve solves a constrained version of the OGTR-All problem in which, in addition to $G^N$ and $S$, we take as input a set of unrooted gene trees that define constraints on the set of binary resolutions of $G^N$. The set of unrooted gene trees used should represent a sample of gene tree topologies supported by the sequence data and can be easily obtained by either computing bootstrap replicates or sampling from the posterior distribution in a Bayesian analysis.

More formally, let $\mathbb{B} = \{B_1, B_2, \ldots, B_b\}$ denote a sample of $b$ unrooted gene trees. Then, we define the cluster set of $\mathbb{B}$ to be: $Cluster(\mathbb{B}) = \bigcup_{i=1}^{b} Cluster(B_i)$. This set of clusters is used to define the constrained set of binary resolutions as follows.

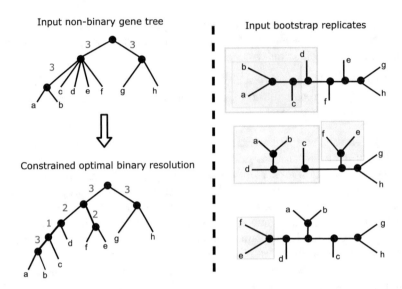

**Fig. 1.** Constrained binary resolutions. Given the (rooted) non-binary gene tree on the top left and the three (unrooted) bootstrap replicate gene trees on the right, the figure shows a possible constrained binary resolution of the non-binary gene tree. Note that each new clade in the binary resolution appears as a cluster in at least one of the three bootstrap replicate trees. These clusters are highlighted using the yellow boxes. Each internal edge in the gene trees is labeled by its branch support (number in red), i.e., the number of bootstrap replicates that support that branch. In this example, the constrained binary resolution happens to be a constrained *optimal* binary resolution since no other constrained binary resolution has higher average branch support. (Color figure online)

**Definition 2 (Constrained binary resolution).** *Given* $\mathbb{B}$ *and a non-binary tree* $T$, *we say that* $T'$ *is a* constrained binary resolution *of* $T$ *(with respect to* $\mathbb{B}$*), if* $T' \in \mathcal{BR}(T)$ *and* $Clade(T') \subseteq Cluster(\mathbb{B})$. *We denote by* $\mathcal{CBR}(T)$ *the set of all constrained binary resolutions of a rooted non-binary tree* $T$.

The idea of a constrained binary resolution is illustrated in Fig. 1. We can now state the constrained optimal gene tree resolution problem.

**Problem 2 (C-OGTR).** *Given* $G^N$, $S$, *and* $\mathbb{B}$, *along with* $P_\Delta$, $P_\Theta$, *and* $P_{loss}$, *the* All Constrained Optimal Gene Tree Resolutions (C-OGTR) *problem is to compute the set* $\mathcal{COR}(G^N)$ *of all optimal constrained binary resolutions of* $G^N$ *such that, for any* $G^B \in \mathcal{COR}(G^N)$, *an MPR of* $G^B$ *and* $S$ *has the smallest reconciliation cost among all gene trees in* $\mathcal{CBR}(G^N)$.

Note: To ensure that a solution always exists to the C-OGTR problem, Tree-Solve includes the original binary gene tree from which $G^N$ is obtained in the set $\mathbb{B}$. This ensures, that a constrained binary resolution of $G^N$ always exists.

We also define a variant of the problem above that only seeks to find a single optimal reconciliation with highest average clade support.

**Problem 3 (C-OGTR-Best).** *Given $G^N$, $S$, and* $\mathbb{B}$, *along with $P_\Delta$, $P_\Theta$, and $P_{loss}$, the* Best Constrained Optimal Gene Tree Resolutions (C-OGTR-Best) *problem is to compute a tree $G^B \in CBR(G^N)$ such that the total number of occurrences in* $\mathbb{B}$ *of all clades in $G^B$ is the largest among all trees in $CBR(G^N)$.*

Note that TreeSolve does not directly solve the *C-OGTR* and *C-OGTR-Best* problems. Rather, for improved efficiency and accuracy, TreeSolve solves variants of these problems where *Cluster*($\mathbb{B}$) is further restricted to only contain those clusters that are present in at least a certain number, *minSup*, of the samples in $\mathbb{B}$, where *minSup* is updated dynamically during the search. Furthermore, TreeSolve maintains ordered lists of binary resolutions at each step sorted by average support value (i.e., by the total number of occurrences in $\mathbb{B}$ of all clades in that binary resolution).

## 3  Algorithmic Overview

Our algorithms for the *C-OGTR* and *C-OGTR-Best* problems leverage the dynamic programming algorithm for the *OGTR-All* problem described in [7]. The primary difference is that the new algorithms limit the possible binary resolutions considered at each non-binary node to those that can be constructed from the clusters available in *Cluster*($\mathbb{B}$). Further technical details are omitted for brevity. Here, we describe how solutions for *C-OGTR/C-OGTR-Best* are used within TreeSolve as part of the larger self-adaptive approach and optimal resolution ordering.

**TreeSolve's Self-adaptive Algorithm.** Note that, despite the restriction on permitted resolutions imposed by $\mathbb{B}$, the total number of constrained optimal binary resolutions can be exponential in the number of non-binary nodes of $G^N$ as well as its maximum out-degree. To address this limitation, TreeSolve employs a novel self-adaptive approach to limit the number of binary resolutions considered at each non-binary node. To describe the self-adaptive approach, we need some additional definitions and notation. We first define an upper bound, denoted $U$, on the total number of binary resolutions considered by TreeSolve during any step in its execution. For example, for all the experimental results presented in the next section, we assigned $U = 25000$. We also define the following:

$$Cluster(\mathbb{B}, minSup) = \{x \in Cluster(\mathbb{B}) \mid x \text{ appears in at least } minSup \text{ trees from } \mathbb{B}\}.$$

Finally, define $N(g, minSup)$ to be the number of distinct binary resolutions of the non-binary node $g \in G^N$ permitted by the cluster set *Cluster*($\mathbb{B}, minSup$). For each non-binary node $g \in G^N$ independently, TreeSolve computes a value for *minSup* for which $N(g, minSup) \leq U$ but $N(g, minSup-1) > U$. This can be accomplished efficiently through a binary-search in the range $[1, |\mathbb{B}|]$. Thus, at each non-binary node of the gene tree, we limit the total number of resolutions considered to at most $U$ of the most highly supported ones.

**Ordering of Binary Resolutions by Average Clade Support.** In addition to its use for limiting the number of possible resolutions at each non-binary node, TreeSolve also uses the upper bound $U$ to bound the total number of resolutions considered at the subtree rooted at each node of the gene tree. In other words, TreeSolve executes a variant of the algorithm for *C-OGTR* that always limits the total number of resolutions of the subtree $G^N(g)$ stored at any node $g$ of the gene tree to $U$. In particular, at each node $g \in G^N$ the algorithm only stores up to the $U$ best (in terms of average clade support) resolutions for the subtree $G^N(g)$ encountered during the search, ordered by their average clade support. We denote this ordered list of the $U$ best resolutions for the subtree $G^N(g)$ by $\mathcal{ORV}(g)$ (for optimal resolution vector). Note that each resolution stored in $\mathcal{ORV}(g)$ also has an associated average clade support value stored along with it. Next, we describe how each $\mathcal{ORV}(\cdot)$ is computed as part of the bottom-up dynamic programming traversal of $G^N$. We first need some additional notation. Given any binary or non-binary node $g \in G^N$, define the set of nearest non-binary descendants of $g$, denoted $\mathcal{N}(g)$, to be $\{h \in V(G^N(g)) \setminus \{g\} \mid h$ is non-binary and no other non-binary nodes exist on the path from $g$ to $h\}$. Note that $\mathcal{N}(g)$ may be empty.

Consider any binary or non-binary node $g \in G^N$. If all nodes in the subtree $G^N(g)$ are binary then there is only one possible resolution (i.e., the current resolution). If $\mathcal{N}(g) = \emptyset$ but $g$ itself is non-binary then we apply the self-adaptive approach described above and compute up to $U$ binary resolutions of $G^N(g)$. These resolutions are then sorted according to decreasing average clade support (based on the trees in $\mathbb{B}$) and stored as $\mathcal{ORV}(g)$. If $\mathcal{N}(g) \neq \emptyset$ and $g$ is binary, then $\mathcal{ORV}(g)$ can be computed by suitably combining the vectors $\mathcal{ORV}(h)$, for each $h \in \mathcal{N}(g)$, already computed in previous steps of the algorithm. Observe that each combination of resolutions from the $\mathcal{ORV}(h)$'s, across all $h \in \mathcal{N}(g)$, yields a permitted resolution for the subtree $G^N(g)$. Since each $\mathcal{ORV}(h)$ is in sorted order and each resolution is associated with its average clade support value, computing the $U$ best resolutions for $G^N(g)$, i.e., computing $\mathcal{ORV}(g)$, can be accomplished by performing a merge-like procedure (from merge sort) on the $\mathcal{ORV}(h)$'s to identify just the $U$ best resolutions for $G^N(g)$. The remaining case, where $\mathcal{N}(g) \neq \emptyset$ and $g$ is non-binary can be handled similarly by considering the sorted list of permitted resolutions for node $g$ together with the $\mathcal{ORV}(h)$'s.

**Computing Only a Single Best Resolution.** By default, TreeSolve computes a sorted list of up to $U$ (where $U = 25000$ in all our experiments) distinct best resolutions of the initial non-binary gene tree. However, in many applications, only a single best estimate of the error-corrected gene tree may be required. Indeed, most existing species-tree-aware methods for microbial gene tree error-correction, including TreeFix-DTL and ecceTERA, only compute a single best gene tree. It is easy to see that the first tree output by TreeSolve corresponds to this best tree, i.e., with highest average clade support. However, if only the best solution was required, TreeSolve could make use of the simpler *C-OGTR-Best* problem formulation, instead of the *C-OGTR* problem as described above. Solving the *C-OGTR-Best* problem is simpler and more efficient than the *C-OGTR* problem (though still potentially exponential). Specifically, to only compute the resolution with highest average clade support, we need not maintain $\mathcal{ORV}(\cdot)$ vectors and only need to save the best resolution corresponding to each subproblem $c(g, s)$.

## 4    Experimental Evaluation

*Simulated and Real Datasets Used in the Analysis.* To evaluate the performance our new approach, we used the large simulated dataset of 2400 gene tree/species tree pairs on 50 taxa used in [2] to evaluate the accuracy of TreeFix-DTL. These 2400 gene trees represent 24 categories (each with 100 gene trees) that capture a wide range of evolutionary scenarios. Specifically, the datasets represent all combinations of (i) low, medium, and high rates of duplication, transfer, and loss events, (ii) four different sequence mutation rates (rates 1, 3, 5 and 10), and (iii) normal (333 amino acids) and short (173 amino acids) sequence lengths; further details on the construction of datasets are available in [2]. For each of the 2400 gene tree/species tree pairs in this dataset, we have available the true (simulated) gene tree and species tree, the reconstructed maximum likelihood gene tree (constructed using RAxML on sequence data simulated down the true gene tree), and 100 bootstrap replicates computed during the execution of RAxML. The 24 categories in this dataset span a wide range of gene tree sizes, event rates, and error rates: specifically, the average leaf set size of the low, medium, and high DTL gene trees are 52.3, 70.4, and 91.3, respectively; the average count of evolutionary events (duplications, transfers, and losses) for the low, medium, and high DTL gene trees are 5.5, 10.6, and, 18.8, respectively, with transfers constituting roughly half of each count; and baseline RAxML error rates (in terms of NRFD, as defined below) ranging from a low of less than 6% to a high of almost 18%.

To further test the scalability and accuracy of TreeSolve on large datasets, we used a dataset of 200 gene tree/species tree pairs on 200 taxa also used in [2]. These 200 gene trees represent 2 distinct categories corresponding to normal sequence length (333 amino acids), a medium rate of DTL, and mutation rates 1 and 5.

In addition to the simulated dataset above, we also used a real biological dataset of over 4700 gene trees from 100 predominantly prokaryotic species [3]. We use this dataset to demonstrate scalability and the impact of applying TreeSolve in practice.

*Experimental Setup.* We evaluated the accuracy and runtime of TreeSolve, TreeFix-DTL, and ecceTERA on each dataset described above. TreeSolve and ecceTERA both take as input a gene tree with support values on its edges, a support cutoff threshold to collapse edges with weak support (thereby producing a non-binary gene tree), and a rooted species tree. In addition, TreeSolve also takes as input the set $\mathbb{B}$ of bootstrap or other samples based on which the gene tree edge support values were computed. TreeFix-DTL takes as input a maximum-likelihood (e.g., RAxML) gene tree, the corresponding sequence alignment, and a rooted species tree. We used default event cost values of 1, 2, and 3, for losses, duplications, and transfers, respectively, for TreeSolve, ecceTERA, and TreeFix-DTL (all three programs use these same event costs by default). The rooted gene trees given as input to TreeSolve and ecceTERA were obtained by rooting each reconstructed RAxML gene tree at an edge that minimized its DTL reconciliation cost. To create the non-binary gene trees for TreeSolve and ecceTERA, we tried two different support cutoff thresholds: 50% and 90%. Note that using higher bootstrap cutoff values results in more non-binary (i.e., more unresolved) gene trees as more edges are collapsed. We observed that both ecceTERA and TreeSolve performed significantly better when the higher cutoff threshold of 90% was used; specifically, the average error-rate in terms of NRFD (defined below) across the 24 baseline simulated datasets decreased from 7.1% to 5.7% for ecceTERA and from 7.7% to 4.85% for TreeSolve. Thus, we fixed the cutoff threshold to 90% for all our experiments with ecceTERA and Tree-Solve, including those with real biological data. For each simulated and real gene tree, 100 bootstrap replicates obtained through RAxML were used to define the corresponding set $\mathbb{B}$ for TreeSolve.

To measure gene tree accuracy, we used the (unrooted) normalized Robinson-Foulds distance (NRFD) against the true gene tree; for any reconstructed gene tree, the NRFD shows the percentage of splits in that gene tree that do not appear in the corresponding true gene tree. Finally, when evaluating the accuracy of TreeSolve, unless otherwise stated, we use only the best (ie., first) resolution computed.

*Basic Statistics on Datasets.* For the 24 baseline simulated datasets, the average leaf set size of the low, medium, and high DTL gene trees was 52.33, 70.37, and 91.26, respectively. Upon collapsing weakly supported edges with the 90% cutoff threshold, we found that the average number of non-binary nodes and average of maximum out-degrees across all 24 baseline simulated datasets were 10.9 and 8.2, respectively, with the highest averages observed to be 15.9 (for sequence length 173, rate-10, high DTL) and 20.58 (for sequence length 173, rate-1, high DTL), respectively.

For the larger 200-taxon simulated datasets, the average number of non-binary nodes was 39.5 and the average of maximum out-degrees was 6.7.

For the real dataset of 4736 gene trees, we found that 4419 became non-binary at a 90% bootstrap cutoff threshold. For these 4419 non-binary gene trees, the average leaf set size was 36.1, the largest leaf set size was 600, and the average number of non-binary nodes and average of maximum out-degrees were 3.35 and 21.14, respectively.

## 4.1   Results

**Simulated Datasets Results.** We first compared the accuracies of TreeSolve, ecceTERA, and TreeFix-DTL on the 24 baseline simulated datasets. These results are shown in Fig. 2. As the figure shows, *TreeSolve* results in significantly more accurate gene tree resolutions than *ecceTERA* in 19 out of the 24 datasets (and across all high DTL datasets), while TreeFix-DTL outperforms both ecceTERA and TreeSolve on all 24 datasets. As we discuss later, this improved accuracy of TreeFix-DTL comes at the expense of orders of magnitude greater running time. The average normalized Robinson-Foulds distances (NRFD) for RAxML, ecceTERA, TreeSolve, and TreeFix-DTL are 7.4%, 3.9%, 3.1%, and 1.86%, respectively, across the 12 normal sequence-length datasets, and 12.5%, 7.6%, 6.6%, and 3.8%, respectively, across the 12 short sequence datasets. As expected, all three species-tree-aware methods were significantly more accurate than the sequence-only method RAxML, and absolute error rates for all four methods were higher for the short (173) sequence length datasets than for the normal (333) length datasets. Interestingly, we observed that while the accuracies of ecceTERA and TreeFix-DTL consistently worsen with increasing DTL rates, the accuracy of TreeSolve is only slightly affected by DTL rates (Fig. 2). As a result, the accuracy of TreeSolve starts to approach that of TreeFix-DTL on the high DTL datasets. Specifically, the average NRFDs across the high-DTL datasets for TreeFix are 2.7% and 5.1% for the normal and short sequence-length datasets, respectively, while for TreeSolve these numbers are only slightly larger at 3.5% and 6.9%, respectively.

**Impact of Gene Tree Size.** To study the impact of tree size on the relative accuracies of the three methods, we used the two simulated datasets of 100 gene tree/species tree pairs each on 200 taxa. As expected, TreeSolve continues to significantly outperform ecceTERA on these larger datasets, with an average NRFD of 3.0% for ecceTERA and only 1.8% for TreeSolve. More significantly, we find that TreeSolve slightly outperforms TreeFix-DTL on these larger trees, with average NRFD of 1.8% to TreeFix-DTL's 1.85%. This is not entirely surprising, since TreeFix-DTL relies on an iterative local search approach that can become less effective as tree size increases. Thus, TreeSolve can be expected to outperform all other methods for larger gene trees.

**Impact of Enumerating Multiple Optimal Resolutions.** Recall that a key feature of TreeSolve is that it can compute and output multiple optimal resolutions, ordered by their average support values. To explore the impact of considering multiple optimal resolutions instead of only using the "best" resolution computed through TreeSolve, we computed the false positive and false negative branch rates for the strict consensus of all multiple optimal resolutions computed by TreeSolve. We found that the strict consensus of all optimal resolutions computed by TreeSolve results in a significantly lower false positive rate compared to just using the "best" TreeSolve gene trees across each of the 24 datasets, with an overall average of 3.45% versus 5.85%, respectively. This suggests that the optimal resolutions computed by TreeSolve can be used to distinguish between correct and incorrect gene tree edges. Unsurprisingly, this improvement in the false positive rate comes at the expense of an increased false negative rate, with the average normalized false negative rate over all 24 datasets being 8.5 for the strict consensus of all multiple optimal resolutions computed by TreeSolve. For brevity, detailed results on individual datasets are omitted from this manuscript.

Error-rates for normal sequence-length baseline datasets

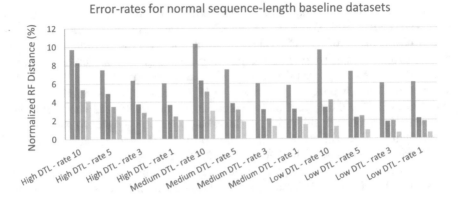

Error-rates for short sequence-length baseline datasets

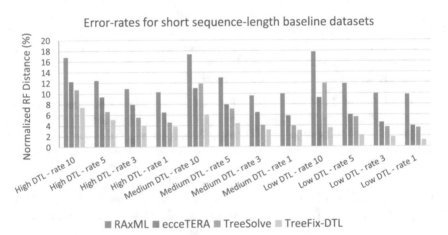

■ RAxML  ■ ecceTERA  ■ TreeSolve  ■ TreeFix-DTL

**Fig. 2.** Accuracy on baseline datasets. Error rates are shown for gene trees inferred using RAxML, ecceTERA, TreeSolve, and TreeFix-DTL on the 12 normal sequence-length (top) and 12 short sequence-length (bottom) simulated datasets.

**Running Time and Scalability.** Both ecceTERA and TreeSolve required only a few seconds per simulated gene tree. Specifically, the average running time of ecceTERA was 2.9 s per tree across the 24 baseline simulated datasets, and for TreeSolve the corresponding average running time was 10.2 s. TreeFix-DTL was far slower than ecceTERA and TreeSolve, requiring an average of over an hour for each of the trees in these 24 baseline simulated datasets. On the larger 200-taxon simulated datasets, ecceTERA and TreeSolve averaged 2.5 s and 82 s per gene tree, respectively. In contrast, TreeFix-DTL required an average of over 10 h per gene tree. Thus, TreeSolve is almost three orders of magnitude faster than TreeFix-DTL on these larger gene trees while also showing better accuracy. All timed runs were executed using a single core on a commodity Macbook Pro laptop with 16 GB of RAM and a 2.3 GHz Intel i9 CPU.

**Results on Real Dataset.** We studied the impact of applying ecceTERA, TreeSolve and TreeFix-DTL on a real biological dataset of over 4736 gene trees from 100 predominantly prokaryotic species [3]. We found that 4419 out of the 4736 gene trees became non-binary at the 90% bootstrap cutoff threshold, and ecceTERA and TreeSolve were thus able to error-correct these 4419 trees. Over all gene trees, ecceTERA had an average running time of 15.1 s and a maximum running time of 317 s. TreeSolve had a slightly larger average running time of 71.6 s and a maximum of 1736 s. Note that the largest gene tree in this dataset has 600 leaves. This demonstrates how TreeSolve can be applied to very large gene trees within minutes.

For the 4419 non-binary gene trees, we found that ecceTERA resulted in an average decrease of 26.4% in the reconciliation cost of the error-corrected gene trees. For TreeSolve, this decrease was a much larger 38.5%. The magnitude of decrease in reconciliation cost is a highly imperfect proxy for gene tree accuracy; still, these numbers suggest that TreeSolve is more effective at error-correcting these real gene trees.

In contrast to ecceTERA and TreeSolve, which executed within minutes on even the largest gene trees, TreeFix-DTL required more than a week of running time on each of the larger gene trees in this dataset.

## 5   Conclusion

In this work, we introduced a new species-tree-aware method, TreeSolve, for error-correcting microbial gene trees. TreeSolve combines new and existing techniques and uses novel algorithms to strike a balance between speed and accuracy. As our extensive experimental analysis demonstrates, TreeSolve significantly outperforms the best existing species-tree-aware methods for microbes in terms of accuracy, speed, or both. TreeSolve is especially effective for error-correction of large gene trees, where it makes it possible to perform speedy error-correction without any compromise on reconstruction accuracy. Furthermore, TreeSolve has the extremely useful ability to compute not just a single best estimate of the error-corrected gene tree but a ranked list of multiple distinct "roughly equally good" candidates. As we show in our experimental study, the resulting ability to aggregate over multiple gene tree candidates helps distinguish between correct and incorrect relationships in an error-corrected gene tree. Overall, TreeSolve has the potential to transform the reconstruction of large microbial gene trees and to increase the robustness of downstream evolutionary inferences by enabling the accounting of gene tree reconstruction uncertainty.

**Funding.** This work was supported in part by NSF awards MCB 1616514 and IES 1615573 to MSB.

# References

1. Bansal, M.S., Alm, E.J., Kellis, M.: Efficient algorithms for the reconciliation problem with gene duplication, horizontal transfer and loss. Bioinformatics **28**(12), 283–291 (2012)
2. Bansal, M.S., Wu, Y.C., Alm, E.J., Kellis, M.: Improved gene tree error correction in the presence of horizontal gene transfer. Bioinformatics **31**(8), 1211–1218 (2015). https://doi.org/10.1093/bioinformatics/btu806
3. David, L.A., Alm, E.J.: Rapid evolutionary innovation during an archaean genetic expansion. Nature **469**, 93–96 (2011)
4. Doyon, J.-P., Scornavacca, C., Gorbunov, K.Y., Szöllősi, G.J., Ranwez, V., Berry, V.: An efficient algorithm for gene/species trees parsimonious reconciliation with losses, duplications and transfers. In: Tannier, E. (ed.) RECOMB-CG 2010. LNCS, vol. 6398, pp. 93–108. Springer, Heidelberg (2010). https://doi.org/10.1007/978-3-642-16181-0_9
5. Jacox, E., Weller, M., Tannier, E., Scornavacca, C.: Resolution and reconciliation of non-binary gene trees with transfers, duplications and losses. Bioinformatics **33**(7), 980 (2017). https://doi.org/10.1093/bioinformatics/btw778. http://dx.doi.org/10.1093/bioinformatics/btw778
6. Kordi, M., Bansal, M.S.: On the complexity of duplication-transfer-loss reconciliation with non-binary gene trees. IEEE/ACM Trans. Comput. Biology Bioinform. **14**(3), 587–599 (2017)
7. Kordi, M., Bansal, M.S.: Exact algorithms for duplication-transfer-loss reconciliation with non-binary gene trees. IEEE/ACM Trans. Comput. Biology Bioinform. **16**(4), 1077–1019 (2019)
8. Li, H., et al.: Treefam: a curated database of phylogenetic trees of animal gene families. Nucleic Acids Res. **34**(Suppl. 1), D572–D580 (2006)
9. Nguyen, T.H., Doyon, J.-P., Pointet, S., Chifolleau, A.-M.A., Ranwez, V., Berry, V.: Accounting for gene tree uncertainties improves gene trees and reconciliation inference. In: Raphael, B., Tang, J. (eds.) WABI 2012. LNCS, vol. 7534, pp. 123–134. Springer, Heidelberg (2012). https://doi.org/10.1007/978-3-642-33122-0_10
10. Rasmussen, M.D., Kellis, M.: A bayesian approach for fast and accurate gene tree reconstruction. Mol. Biol. Evol. **28**(1), 273–290 (2011). https://doi.org/10.1093/molbev/msq189
11. Scornavacca, C., Jacox, E., Szollosi, G.J.: Joint amalgamation of most parsimonious reconciled gene trees. Bioinformatics **31**(6), 841 (2015). https://doi.org/10.1093/bioinformatics/btu728
12. Sjostrand, J., Tofigh, A., Daubin, V., Arvestad, L., Sennblad, B., Lagergren, J.: A bayesian method for analyzing lateral gene transfer. Syst. Biol. **63**(3), 409–420 (2014). https://doi.org/10.1093/sysbio/syu007
13. Stolzer, M., Lai, H., Xu, M., Sathaye, D., Vernot, B., Durand, D.: Inferring duplications, losses, transfers and incomplete lineage sorting with nonbinary species trees. Bioinformatics **28**(18), 409–415 (2012)
14. Szollosi, G.J., Boussau, B., Abby, S.S., Tannier, E., Daubin, V.: Phylogenetic modeling of lateral gene transfer reconstructs the pattern and relative timing of speciations. Proc. Natl. Acad. Sci. USA **109**(43), 17513–17518 (2012). https://doi.org/10.1073/pnas.1202997109
15. Szollosi, G.J., Tannier, E., Lartillot, N., Daubin, V.: Lateral gene transfer from the dead. Syst. Biol. **62**(3), 386 (2013). https://doi.org/10.1093/sysbio/syt003. http://dx.doi.org/10.1093/sysbio/syt003

16. Tofigh, A., Hallett, M.T., Lagergren, J.: Simultaneous identification of duplications and lateral gene transfers. IEEE/ACM Trans. Comput. Biol. Bioinform. **8**(2), 517–535 (2011)
17. Zheng, Y., Zhang, L.: Reconciliation with non-binary gene trees revisited. In: Sharan, R. (ed.) RECOMB 2014. LNCS, vol. 8394, pp. 418–432. Springer, Cham (2014). https://doi.org/10.1007/978-3-319-05269-4_33

# RNA-Seq and Other Biological Processes

# Time Series Adjustment Enhancement of Hierarchical Modeling of *Arabidopsis Thaliana* Gene Interactions

Edward E. Allen[1], John Farrell[2], Alexandria F. Harkey[3], David J. John[2(✉)], Gloria Muday[3], James L. Norris[1], and Bo Wu[1]

[1] Mathematics and Statistics, Wake Forest University, Winston-Salem, NC 27109, USA
{allene,norris,wub18}@wfu.edu
[2] Computer Science, Wake Forest University, Winston-Salem, NC, USA
{farrjt17,djj}@wfu.edu
[3] Biology, Wake Forest University, Winston-Salem, NC, USA
{harka14,muday}@wfu.edu

**Abstract.** Network models of gene interactions, using time course gene transcript abundance data, are computationally created using a genetic algorithm designed to incorporate hierarchical Bayesian methods with time series adjustments. The posterior probabilities of interaction between pairs of genes are based on likelihoods of directed acyclic graphs. This algorithm is applied to transcript abundance data collected from *Arabidopsis thaliana* genes. This study extends the underlying statistical and mathematical theory of the Norris-Patton likelihood by including time series adjustments.

**Keywords:** Gene interaction network modeling · Bioinformatics · Genetic algorithms · Bayesian methods · Time series adjustment

## 1 Introduction

Cell signaling is accomplished via networks of transcriptional changes that lead to synthesis of distinct sets of proteins, which cause changes in growth, development, or metabolism. Treatments that elevate levels of hormones result in cascades of changes in gene expression driven by activation and synthesis of transcription factors which are required to turn on downstream genes. One approach to model these gene regulatory networks is to collect measurements of changes in abundance of gene transcripts across a time course. The expression of a gene encoding a transcriptional activator or repressor protein may signal to the next gene to either turn on or turn off downstream genes and their encoded proteins. Thus, time course transcriptomic data sets contain important information about how genes drive these changes in biological networks. Yet genome-wide transcript abundance assays examine tens of thousands of genes so identification of patterns or networks within these large data sets is difficult. It is also

© Springer Nature Switzerland AG 2020
C. Martín-Vide et al. (Eds.): AlCoB 2020, LNBI 12099, pp. 143–154, 2020.
https://doi.org/10.1007/978-3-030-42266-0_11

critical to filter the meaningful transcript changes in these data sets to remove genes whose responses are not above background or that are dissimilar due to biological or technical variation. Yet even though the bioinformatics community has developed statistical methods to filter the data [9], additional approaches are needed to identify the networks and patterns in these large data sets.

An important modern approach to statistical modeling includes Bayesian techniques involving likelihoods and posterior probabilities. Here, we extend our previous work on this problem by incorporating time series adjustments in the computation of Bayesian likelihoods. We apply this method to time course data generated in response to treatments that elevate the levels of the hormone ethylene in *Arabidopsis thaliana*. We take advantage of a previously published genome-wide transcriptional data set [9], subjected to rigorous filtering and from which all the genes predicted to encode transcription factors have been identified. The goal is to predict gene regulatory networks that control time-matched developmental changes.

The results in this paper are novel for several reasons. First, the methods use the hierarchical nature of the data sets. For example, replicate data are not averaged. Rather, the method constructs a model over all of the data that uses each replicate as a source of information. The assumption is that at each level of the hierarchy there are commonalities in the data and parameters. Thus, the replicate data is not independent. Second, the addition of time series adjustment to improve the independence of the model's residuals gives these techniques stronger statistical foundations. Third, the combination of Bayesian model averaging with a cutting edge genetic algorithm provides rigorous estimates of posterior probabilities for edges. These computational modeling algorithms are derived using rigorous mathematical and statistical techniques and are computationally efficient. The models produced are easily understandable.

Many different techniques for modeling non-hierarchical data using gene expression data have been proposed. An excellent recent survey on this subject was given by Emily [4]. There are many techniques for modeling gene and protein networks–with various different properties–available in the literature. Our technique in this paper is a Bayesian regression type method. Variations of Bayesian modeling can be found in [7,11,19]. Other methods that use types of regression include [2,21] which focus on logistic regression techniques, and [22,23] which use Poisson regression. Other approaches to modeling these types of problems include differential equations [1] and Boolean modeling [14].

This Bayesian likelihood computational algorithm incorporates additional important features from earlier versions. Earlier variations included computing posterior probabilities for a single replicate [11] and for multiple replicates with both hierarchical [18] and independent [17] structures. Over the course of this research, the search procedure has changed from Metropolis Hastings to genetic algorithms. Genetic algorithms' execution times are typically polynomial rather than the doubly exponential execution time, in terms of the numbers of time points and genes, of Metropolis Hastings.

This variation also uses a Bayesian version of the *Cross generational elitist selection, Heterogeneous recombination, Cataclysmic mutation* algorithm (CHC) [6]. Genetic algorithms are motivated by the operators of selection, crossover, and mutation. The CHC variation does not allow the crossover of similar parents. Once the population becomes too homogeneous, then a cataclysmic mutation event regenerates the population from the current *most fit* parents. The Bayesian CHC (BCHC) implemented in this paper uses a hierarchical statistical construct (the Norris-Patton Likelihood) as the fitness function.

The hormone ethylene (ACC) is known to activate root growth in *Arabidopsis thaliana* [9]. Transcription factors (TFs) are cellular proteins that bind to DNA to turn genes either *on* (activation) or off (repression). Developmental changes are controlled by these genes. The data set used in this modeling process was the complete set of abundance levels of the twenty-six TFs believed/known to be involved in the activation of the growth of roots at eight time points after treatment with the ethylene precursor ACC [9]. Here, constructing an appropriate network model has potential agricultural applications in that it should lead to more complex understandings of root development.

## 2   Mathematical and Statistical Preliminaries

Three network modeling paradigms are generally considered in the literature: *cotemporal, next state one step* and *next state one and two steps*. A next state one step model predicts the transcript abundance relationships between genes at time $j$ based on the transcript abundance at time $j-1$. In this paper, we will only consider next state one step models; for simplicity, we will refer to next state one step as next state. The time series adjusted (tsa) next state models are an amalgamation of next state modeling with standard time series adjustments [12]. The time series adjustment methodology makes the residuals (i.e., the estimated error terms) more independent.

A *directed graph* $G = (V, E)$ consists of a pair of collections: $V$ a set of vertices (or nodes); and, $E$ a collection of directed edges between pairs of vertices. A cycle is a sequence $v_1, e_1, v_2, e_2, \ldots, v_{n-1}, e_{n-1}, v_n = v_1$ where $v_i \in V$ and $e_j \in E$ is a directed edge from vertex $v_{j-1}$ to vertex $v_j$. Directed acyclic graphs (DAGs) do not contain cycles. An example of a DAG is given in Fig. 1. In this modeling algorithm, DAGs form the mathematical foundation of our computational approach. The vertices of a DAG represent genes and the directed edges are one-way relationships between pairs of vertices. When there is a directed edge from $v_i$ to $v_j$, then $v_i$ is a parent of $v_j$ and $v_j$ is a child of $v_i$.

For any DAG $D$ with vertex set $V = \{v_1, v_2, \cdots, v_n\}$, the vertices can be topologically sorted. This gives a total order $>$ on $V$ such that if $v_i$ is an ancestor of $v_j$, meaning that there is a directed path from $v_i$ to $v_j$, then $v_i < v_j$. Without loss of generality, lets assume that $v_i < v_{i+1}$ for $1 \leq i \leq n-1$.

Conditional probability gives that for any two events $A$ and $B$, the probability

$$P(A \text{ and } B) = P(A) \, P(B|A) = P(B) \, P(A|B).$$

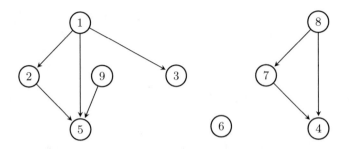

**Fig. 1.** A directed acyclic graph. Gene 1 affects genes 2, 3 and 5 but not genes 4, 6, 7, 8 and 9. Genes 1, 6 and 8 are not affected by any other gene.

Similarly, the density function $f$ for two continuous variables $y_1$ and $y_2$ is

$$f(y_1 \text{ and } y_2) = f(y_1, y_2) = f(y_1)\, f(y_2|y_1) = f(y_2)\, f(y_1|y_2)$$

Recursively, using the order $<$ implied by topologically sorting the DAGs on the set of continuous variable $Y = \{y_1, y_2, \ldots, y_n\}$ (i.e., $y_i < y_j$ if and only if $v_i < v_j$), gives

$$f(y_1, y_2, y_3, \ldots, y_n) = f(y_1)\, f(y_2|y_1)\, f(y_3|y_1, y_2) \ldots f(y_n|y_1, y_2, \ldots, y_{n-1}).$$

Specific for a particular DAG $D$, let $y_1$ be the gene that cannot have any parents. Let $y_2$ be the gene that can have at most parent $y_1$. Similarly, let $y_h$ be the gene that can have parents from the collection $\{y_1, \cdots, y_{h-1}\}$. Therefore, if we let $y_i$ represent the data of child $i$ for all of the $r$ replicates, we have for $D$

$$f(y_1, y_2, \ldots, y_n | D) = f(y_1|D) f(y_2|y_1, D) f(y_3|y_1, y_2, D) \cdots f(y_k|y_1, \cdots y_{n-1}, D)$$

## 3    Time Series Adjustment

Statistical regression models of response (child) data from predictors (parents) data over time nearly always have correlated residuals over time. This is usually due to the remaining influence of the previous time's response data. In complicated modeling situations (e.g., like ours where we need to obtain closed form likelihoods of DAGs within a hierarchical structure in order to produce posterior probabilities of edges), it is common to derive results as if there were non-correlated residuals, as we have done in previous work. Our previous work has shown utility both for simulated and biological data, but we now rigorously incorporate a time series adjustment into our model. This should result in substantially less correlated residuals and thus more accurate likelihoods for the DAGs. Since these likelihoods are the foundations for the edges' estimated posterior probabilities, these estimates should also be improved.

Our time series adjustment is an integer autoregressive adjustment of order 1 in the commonly used family of Markov conditioning. It is a version of Kedem's and Fokianos' autoregressive model [12, page 184]. In our setting, this simply adds the child's data at the previous time as an additional regressor for the child's data at the current time. Thus, much of the child's data at the previous time's influence would be *regressed out* leaving less correlated, closer to independent, residuals from one time to the next.

## 4   Next State Time Series Adjustment Computation

For each $h$, with $1 \leq h \leq n$, $f(y_h|y_1, y_2, \ldots, y_{h-1}, D)$ gives the density of $y_h$ given $y_h$'s parent's data for DAG $D$. Now, let $_iy_c$ be the data vector of any given child $c$ from the $i^{th}$ replicate. The vector $_iy_c$ has dimension $t$, the number utilized time points in the child $c$ data set for a given replicate $i$. The symbol $_ix_c$ is the $t \times k_c$ *regressor* matrix for $_iy_c$. For next state with time series adjustment, $t$ is the number of time points per replicate minus one since at time 1, the child data has no last previous parent data nor last previous child (tsa) data–so, the utilized child data starts at time 2. The value of $k_c$ is the number of parents of $c$ plus two since $_ix_c$ has a separate column for each of its parent's data at the previous time, a column of 1's for the intercept, and a column of the child's data at the previous time (the time series adjustment). A $k_c$ dimensional slope vector for child $c$'s regressors is $_i\beta_c$. The common within replicate residual variance of child $c$ is $\sigma_c^2$.

Assumptions which detail the hierarchical structure include that for a given $_i\beta_c$ and $\sigma_c^2$, each $_iy_c$ is independent and normally distributed, and therefore $_iy_c|_i\beta_c\sigma_c^2 \sim N_t(_ix_c \, _i\beta_c, \sigma_c^2 I)$. Note the $_iy_c$ (child) response and the underlying regression structure of the product of the $_ix_c$ matrix and the $_i\beta_c$ vector. We have $_i\beta_c|\sigma_c^2 \sim N_{k_c}(0, g\sigma_c^2(\bar{x}_c^T \bar{x}_c)^{-1})$ and $\sigma_c^2 \sim$ Inverse-gamma$(v_o/2, v_o\sigma_o^2/2)$. With these assumptions, we have the following result that gives the mathematical and statistical computation of the Norris-Patton likelihood (NPL), $L = f(y_1, \ldots, y_n|D)$, as the value of the density/likelihood function for $D$.

**Theorem 1.** *The closed form solution of the likelihood assuming a hierarchical structure among replicates is* $\prod_{c=1}^{n} f(y_c|parents \ of \ y_c, D)$ *where $n$ is the number of genes and $D$ is a DAG. Also,*

$$f(y_c|parents \ of \ y_c, D) = (2\pi)^{-\frac{rt}{2}} \left(\frac{1}{2}\right)^{-\frac{(rt+v_0)}{2}} g^{-\frac{rk_c}{2}} \Gamma(v_0\sigma_0^2/2)\frac{\Gamma[(rt+v_0)/2]}{\Gamma(v_0/2)}$$

$$\times \frac{|\bar{x}_c^T \bar{x}_c|^{\frac{r}{2}}}{\prod_{i=1}^{r}|_ix_c^T \, _ix_c + \bar{x}_c^T \bar{x}_c\left(\frac{1}{g}\right)|^{\frac{1}{2}}}$$

$$\times \left[v_0\sigma_0^2 + \sum_{i=1}^{r} {_iy_c^T} \, _iy_c - ({_ix_c^T} \, _iy_c)^T[({_ix_c^T} \, _ix_c + \bar{x}^T \, _i\bar{x}\frac{1}{g})^{-1}]^T \, _ix_c^T \, _iy_c\right]^{-\frac{rt+v_0}{2}}$$

The proof of Theorem 1 uses the following lemmas whose computation can be found in [16] (a thesis from our research group). We include the proof of Lemma 2 to show how the computation of the likelihood includes the slope parameters $_i\beta_c$ of each of the replicates separately.

**Lemma 2.** *The contribution of child $c$ to the likelihood to DAG $D$ is*

$$f(y_c \mid D) = f(_1y_c, \ldots, _ry_c \mid D)$$
$$= \int_{\sigma_c^2} [f(_1y_c \mid \sigma_c^2) \cdots f(_ry_c \mid \sigma_c^2)] \, f(\sigma_c^2) \, d\sigma_c^2$$

*Proof.* Using integration, we have

$$f(_1y_c \mid D) = f(_1y_c, \ldots, _ry_c \mid D)$$
$$= \int_{\sigma_c^2} \int_{_r\beta_c} \cdots \int_{_1\beta_c} f(_1y_c, \ldots, _ry_c, _1\beta_c, \ldots, _r\beta_c, \sigma_c^2 \mid D) d_1\beta_c \cdots d_r\beta_c \, d\sigma_c^2$$
$$= \int_{\sigma_c^2} \int_{_r\beta_c} \cdots \int_{_1\beta_c} f(_1y_c, \ldots, _ry_c, \mid _1\beta_c, \ldots, _r\beta_c\sigma_c^2, D)$$
$$\times f(_1\beta_c, \ldots, _r\beta_c \mid \sigma_c^2) \, f(\sigma_c^2) \, d_1\beta_c \cdots d_r\beta_c \, d\sigma_c^2$$
$$= \int_{\sigma_c^2} [f(_1y_c \mid \sigma_c^2) \cdots f(_ry_c \mid \sigma_c^2)] \, f(\sigma_c^2) \, d\sigma_c^2 \qquad \square$$

Letting $|M|$ denote the determinant of the matrix $M$, we have the following:

**Lemma 3.** *For a given replicate $i$ and letting $exp(x)$ represent the exponential function $e^x$, we have*

$$f(_iy_c \mid \sigma_c^2)$$
$$= (2\pi\sigma_c^2)^{-\frac{t}{2}} \mid g\sigma_c^2(\bar{x}_c^T\bar{x}_c)^{-1} \mid^{-\frac{1}{2}} \mid _iA_c \mid^{\frac{1}{2}} exp\left(-\frac{1}{2}\left[\frac{1}{\sigma_c^2}\, _iy_c - _im_c^T \, _iA_c^{-1} \, _im_c\right]\right)$$

*where*

$$_iA_c^{-1} = \frac{1}{\sigma_c^2}\left(_ix_c^T \, _ix_c + \bar{x}_c^T\bar{x}_c\left(\frac{1}{g}\right)\right)$$

*and*

$$_im_c = \left(_ix_c^T \, _ix_c + \bar{x}_c^T\bar{x}_c\left(\frac{1}{g}\right)\right) _ix_c^T \, _iy_c.$$

Extending Lemma 2 to the product of density functions used in Lemma 1, we have:

**Lemma 4**

$$f(_1y_c \mid \sigma_c^2) f(_2y_c \mid \sigma_c^2) \cdots f(_ry_c \mid \sigma_c^2)$$
$$= (2\pi)^{-\frac{rt}{2}} (\tau_c)^{\frac{rt}{2}} (g)^{-\frac{rk_c}{2}} \frac{\mid \bar{x}_c^T\bar{x}_c \mid^{\frac{r}{2}}}{\prod_{i=1}^r \mid _ix_c^T \, _ix_c + \bar{x}_c^T\bar{x}_c\frac{1}{g} \mid^{\frac{1}{2}}}$$
$$\times exp\left(-\frac{1}{2}\tau_c \sum_{i=1}^r \left[_iy_c^T \, _iy_c - (_ix_c^T \, _iy_c)^T [(_ix_c^T \, _ix_c + \bar{x}_c^T\bar{x}_c\left(\frac{1}{g}\right))^{-1}]^T \, _ix_c^T \, _iy_c\right]\right)$$

Note that $g$, $v_0$ and $\sigma_c^2$ are positive free parameters. In our modeling algorithm, we set $g = v_0 = \sigma_c^2 = 1$. The use of the time series adjusted next state Norris-Patton likelihood, along with a tailor-made genetic algorithm and Bayesian model averaging, allows for the rigorous estimation of posterior probabilities for all gene pair interactions.

```
 1: procedure TBCHC
 2:     t ← 0
 3:     Archive ← {}
 4:     multi-step initialization of 400 DAG(s) for P(0)
 5:     indicator ← 50
 6:     while t < 600 do
 7:         t ← t + 1
 8:         X ← randomly reorderP(t−1)
 9:         Y ← {}
10:         for all parent pairs (X[2i], X[2i+1]) do
11:             if parent pair (X[2i], X[2i+1]) are dissimilar then
12:                 Y ← Y ∪ {crossover-repair (X[2i], X[2i+1])}
13:             end if
14:         end for
15:         indicator ← indicator − (|P(t−1)| − |Y|)
16:         P(t) ← NPL fittest |P(t−1)| of P(t−1) ∪ Y
17:         if indicator < 0 then
18:             P(t) ← cataclysm(P(t))
19:             indicator ← 50
20:         end if
21:         Archive ← Archive ∪ P(t)
22:     end while
23:     return Archive
24: end procedure
```

**Fig. 2.** The TBCHC genetic algorithm searches for and returns an archive of unique DAGs (lines 3, 21 and 23). After applying Bayesian model averaging to the archive, the gene interaction model is formed. The initial population consists of 400 DAGs. The time series adjustment is applied in finding the fittest DAGs (line 16). The variable *indicator* triggers cataclysmic mutation (lines 17–19).

## 5    Genetic Algorithms

Simply put, a genetic algorithm (GA) takes the current population and produces the next generation using the operations of *selection*, *crossover*, and *mutation* [15]. Individuals (i.e., DAGS) are automatically moved to the next generation with preference given to those with the higher likelihoods (the *elitist* strategy). The first population must be initialized. The genetic algorithm terminates after a specified number of iterations.

The TBCHC genetic algorithm is an extension of BCH [13] which was heavily influenced by the CHC [5]. The TBCHC fitness function includes the next state time series adjustment. The TBCHC operators of *selection, crossover, mutation,* and *repair* will be discussed in the following paragraphs.

The population of each generation consists of a fixed number of DAGs. Each DAG represents gene relationships. The genetic algorithm's aim is to move from the current population of DAGs to a new generation where the overall quality improves (as measured by the Norris-Patton likelihood). The elitist strategy only moves the top 10% of DAGs from the current generation to the next and the balance is filled by crossover. As TBCHC iterates, all distinct DAGs are archived. The final gene interaction model is produced from this archived collection.

Generally, the selection operator chooses which members of the current population can potentially contribute children to the next generation. In Fig. 2 selection is accomplished through a random pairing of all parents in the current population (lines 8–10). By assuming prior probabilities for the DAG, the likelihood of a given DAG $D$ is proportional to the $D$'s NPL [3]. Thus, the fitness of a candidate $D$ can be computed using the NPL.

The crossover operator (line 12) exchanges genetic information (i.e., directed edges) between two parents producing two new offspring. The edges chosen to be exchanged are chosen randomly. There is one caveat: if the two parents are too similar–determined by the Hamming distance between them then the two selected parent DAGs are not allowed to produce offspring (line 11). In a simple genetic algorithm, all selected parents are allowed to produce offspring. This TBCHC prohibition of mating by similar parents may result in fewer DAGS in the next population than in the current population. Since the modeling process is based on DAGs, if the crossover operator introduces a cycle in the offspring, a repair operator is applied. Selection and crossover are used exclusively in TBCHC until the population becomes too similar. At that point, cataclysmic mutation (line 17) is applied to reset the population by creating a new population of DAGs from the top 10% NPL DAGs.

There are no known techniques for assigning the optimum values to the genetic algorithm parameters. However, experience and the literature give general criterion for appropriate values. Still, values are often determined on a case by case basis. The TBCHC algorithm parameters include the following: 20 parallel executions each with 600 generations; the number of initial DAGS is 400; the crossover probability is 0.30; and, the number of parents of any given node is limited to 3. Cataclysmic mutation causes the population of DAGs to be replaced by DAGs generated by crossover and mutation on the top 10% of the population to restore the candidate class to 400.

This TBCHC algorithm is implemented in *python 3.0* using the *NetworkX* [8] and *dispy* packages [20].

## 6    Gene Interaction Model and Bayesian Model Averaging

It is important to realize that each directed edge in the model is labeled by a number in the interval [0, 1] indicating the posterior Bayesian probability that the associated relationship exists in the biological network. Using Bayesian statistics,

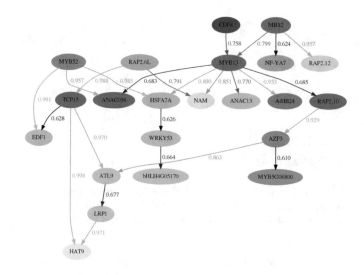

(a) NS TSA Gene Interaction Model.

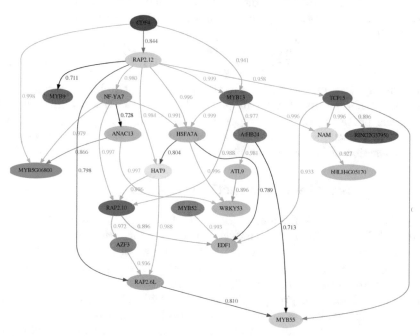

(b) NS (without TSA) Gene Interaction Model.

**Fig. 3.** (a) is the next state time series adjusted model for ACC26 and its analysis. The numerical label on the directed edges is the posterior probability. For clarity, only edges with posterior probability greater than 0.3 are indicated. As a consequence, four genes (MYB55, MYB9, BYB93, and RING2G37950) are not shown in (a) and four genes (ANAC058, LRP1, MBS2, and MYB93) are not shown in (b).

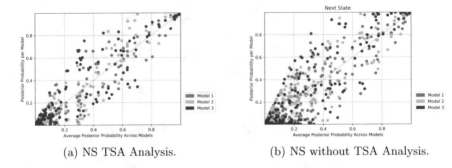

(a) NS TSA Analysis.  (b) NS without TSA Analysis.

**Fig. 4.** Across three independently generated similar gene interaction models, plotting the edges' average posterior probabilities versus the individual edge posterior probabilities provides a consistency analysis of the models given in Fig. 3.

these probabilities are estimated by a weighted sum over all of the models found in the archive $AR$. With $L = f(y_1, \ldots, y_n | D)$ the NPL of data $y_1, \ldots, y_n$, and $\chi_D(e) = 1$ if $e$ is a directed edge in DAG $D$ and $\chi_D(e) = 0$ otherwise, the posterior probability of an edge $e$ is computed by the Bayesian model averaging formula [10]

$$\frac{\sum_{D \in AR} \chi_D(e) f(y_1, y_2, \ldots, y_n | D)}{\sum_{D \in AR} f(y_1, y_2, \ldots, y_n | D)},$$

which simply and appropriately weights each visited DAG $D$ according to its likelihood. This methodology requires equally likely priors since in such a situation the posterior for $D$ is proportional its likelihood [3]. In order for this estimate to reflect its true value, it is necessary that $AR$ contain a large and varied collection of DAGs of high likelihood.

## 7    Next State Gene Interaction Models

Using the transcript abundance data for 26 *Arabidopsis thaliana* genes stimulated by ACC, gene interaction models for a next state with and without time series adjustment were computationally created, shown in Fig. 3. Each edge is labeled by its posterior probability. Figure 4 provides comparisons of three similar models to those given in Fig. 3. Figure 4(a) shows a stronger and tighter distribution of posterior probabilities than Fig. 4(b). There is significant agreement across the models for average posterior probabilities exceeding 0.8 and less than 0.2. However, for average posterior probabilities with values greater than 0.2 and less than 0.8 there is a great deal of variance, which reflects the lack of a strong posterior probability over this range.

## 8    Conclusion and Further Considerations

A typical underlying assumption of statistical analysis is that the residuals are independent [3, page 737]. It is well understood, however, that the residuals associated with time course data are not usually independent. By incorporating time series adjustments into the modeling process, the residuals' independence is much improved; thus, yielding a less approximated, more accurate likelihood function.

The continuation of this research includes four tasks. First, the computational networks have been sent to the Muday lab for biological investigation, confirmation and interpretation. Second, in this paper, we investigated the enhancement of times series adjustment on a next state one step model. There are two other *time paradigms*, next state one and two steps and cotemporal, each of which has a time series adjustment analogue and a corresponding Norris-Patton likelihood. Comparing and contrasting the computational results of these three distinct modeling methods–as well as their biological interpretations–are important in understanding the gene interaction models developed using this methodology. Third, we will further consider higher order autoregressive adjustment to continue improving the independence of the residuals. Fourth, effort is underway to implement nonuniform priors in the modeling techniques. This would permit construction of gene interaction models that reflect relationships found in the literature.

**Acknowledgments.** The authors thank the National Science Foundation for their support with a grant, NSF#1716279. John Farrell thanks Wake Forest University for support as a Wake Forest Fellow for Summer 2019.

# References

1. Cao, J., Qi, X., Zhao, H.: Modeling gene regulation networks using ordinary differential equations. In: Next Generation Microarray Bioinformatics, Methods in Molecular Biology, vol. 802, pp. 185–197. Springer (2012). https://doi.org/10.1007/978-1-61779-400-1_12
2. Cordell, H.: Detecting gene-gene interactions that underlie human diseases. Nat. Rev. Genet. **10**(2), 392–404 (2002)
3. DeGroot, M.H., Schervish, M.J.: Probability and Statistics, 4th edn. Addison-Wesley, Boston (2012)
4. Emily, M.: A survey of statistical methods for gene-gene interaction in case-control genome-wide association studies. J. Soc. Fr. Stat. **159**(1), 27–67 (2018)
5. Eschelman, L.J.: The CHC adaptive search algorithm: how to have safe search when engaging in nontraditional genetic recombination. In: Rawlins, G.J.E. (ed.) Foundations of Genetic Algorithms, pp. 265–283. Morghan Kaufmann, Burlington (1991)
6. Eshelman, L.J.: Genetic algorithms. In: Bäck, T., Fogel, D.B., Michalewicz, T. (eds.) Evolutionary Computation 1 - Basic Algorithms and Operators, Chapter 8, vol. 1, pp. 64–80. Institute of Physics Publishing, Bristol (2000)
7. Friedman, N., Linial, M., Nachman, I., Pe'er, D.: Using Bayesian networks to analyze expression data. J. Comput. Biol. **7**(3), 601–620 (2000). https://doi.org/10.1186/gb-2004-5-12-r100
8. Hagberg, A.A., Schult, D.A., Swart, P.J.: Exploring network structure, dynamics, and function using NetworkX. In: Varoquaux, G., Vaught, T., Millman, J. (eds.) Proceedings of the 7th Python in Science Conference, pp. 11–15, Pasadena (2008)

9. Harkey, A.F., et al.: Identification of transcriptional and receptor networks that control root responses to ethylene. Plant Physiol. **176**(3), 2095–2118 (2018). https://doi.org/10.1104/pp.17.00907. http://www.plantphysiol.org/content/176/3/2095

10. Hoeting, J.A., Madigan, D., Raftery, A.E., Volinsky, C.T.: Bayesian model averaging: a tutorial (with comments by M. Clyde, David Draper and E.I. George, and a rejoinder by the authors). Stat. Sci. **14**(4), 382–417 (1999)

11. John, D.J., Fetrow, J.S., Norris, J.L.: Continuous cotemporal probabilistic modeling of systems biology networks from sparse data. IEEE/ACM Trans. Comput. Biol. Bioinform. **8**(5), 1208–1222 (2011). https://doi.org/10.1109/TCBB.2010.95

12. Kedem, B., Fokianos, K.: Regression Models for Time Series Analysis. Wiley, Hoboken (2002)

13. LaPointe, B.A., et al.: A BCHC genetic algorithm model of cotemporal hierarchical Arabidopsis thaliana gene interactions. In: Proceedings of 2018 IEEE International Conference on Bioinformatics and Biomedicine (BIBM), pp. 2701–2708 (2018)

14. Liang, J., Han, J.: Stochastic Boolean networks: an efficient approach to modeling gene regulatory networks. BMC Syst. Biol. **6**(113), 1–20 (2012). http://www.biomedcentral.com/1752-0509/6/113

15. Mitchell, M.: An Introduction to Genetic Algorithms. MIT Press, Cambridge (1998)

16. Patton, K.L.: Bayesian interaction and associated networks from multiple replicates of sparse time-course data. Master's thesis, Wake Forest University, Department of Mathematics (May 2012)

17. Patton, K.L., John, D.J., Norris, J.L.: Bayesian probabilistic network modeling from multiple independent replicates. BMC Bioinform. **13**(Supplement 9), 1–13 (2012)

18. Patton, K.L., John, D.J., Norris, J.L., Lewis, D., Muday, G.: Hierarchical Bayesian system network modeling of multiple related replicates. BMC Bioinform. **7**, 803–812 (2013)

19. Pe'er, D.: Bayesian network analysis of signaling networks: a primer. Sci. STKE **2005**, 1–12 (2005)

20. Pemmasani, G.: dispy: distributed and parallel computing with/for python (2016). http://dispy.sourceforge.net

21. Purcell, S., et al.: PLINK: a toolset for whole-genome association and population-based linkage analysis. Am. J. Hum. Genet. **81**, 559–575 (2007)

22. Wan, X., et al.: BOOST: a fast approach to detecting gene-gene interactions in disease data. Am. J. Hum. Genet. **87**, 325–340 (2010)

23. Yung, L.S., Yang, C., Wan, X., Yu, W.: GBOOST: a GPU-based tool for detecting gene-gene interactions in genome-wide case control studies. Bioinformatics **27**(9), 1309–1310 (2011)

# BESTox: A Convolutional Neural Network Regression Model Based on Binary-Encoded SMILES for Acute Oral Toxicity Prediction of Chemical Compounds

Jiarui Chen[✉], Hong-Hin Cheong, and Shirley Weng In Siu

Department of Computer and Information Science,
University of Macau, Avenida da Universidade, Taipa, Macau SAR, China
{mb85409,mb85514,shirleysiu}@um.edu.mo

**Abstract.** Compound toxicity prediction is a very challenging and critical task in the drug discovery and design field. Traditionally, cell or animal-based experiments are required to confirm the acute oral toxicity of chemical compounds. However, these methods are often restricted by availability of experimental facilities, long experimentation time, and high cost. In this paper, we propose a novel convolutional neural network regression model, named BESTox, to predict the acute oral toxicity ($LD_{50}$) of chemical compounds. This model learns the compositional and chemical properties of compounds from their two-dimensional binary matrices. Each matrix encodes the occurrences of certain atom types, number of bonded hydrogens, atom charge, valence, ring, degree, aromaticity, chirality, and hybridization along the SMILES string of a given compound. In a benchmark experiment using a dataset of 7413 observations (train/test 5931/1482), BESTox achieved a squared correlation coefficient ($R^2$) of 0.619, root-mean-squared error ($RMSE$) of 0.603, and mean absolute error ($MAE$) of 0.433. Despite of the use of a shallow model architecture and simple molecular descriptors, our method performs comparably against two recently published models.

**Keywords:** Drug design · Machine learning · Acute oral toxicity · Toxicity prediction · SMILES · Convolutional neural network

## 1 Introduction

Measuring the chemical and physiological properties of chemical compounds are fundamental tasks in biomedical research and drug discovery [19]. The basic idea of modern drug design is to search chemical compounds with desired affinity, potency, and efficacy against the biological target that is relevant to the disease of interest. However, not only that there are tens of thousands known chemical compounds existed in nature, but many more artificial chemical compounds are

© Springer Nature Switzerland AG 2020
C. Martín-Vide et al. (Eds.): AlCoB 2020, LNBI 12099, pp. 155–166, 2020.
https://doi.org/10.1007/978-3-030-42266-0_12

being produced each year [9]. Thus, the modern drug discovery pipeline is focused on narrowing down the scope of the chemical space where good drug candidates are [7,11]. Potential lead compounds will be subjected to further experimental validation on their pharmacodynamics and pharmacokinetic (PD/PK) properties [2,14]; the latter includes absorption, distribution, metabolism, excretion, and toxicity (ADME/T) measurements. Traditionally, chemists and biologists conduct cell-based or animal-based experiments to measure the PD/PK properties of these compounds and their actual biological effects *in vivo*. However, these experiments are not only high cost in terms of both time and money, the experiments that involve animal testings are increasingly subjected to concerns from ethical perspectives [1].

Among all measured properties, toxicity of a compound is the most important one which must be confirmed before approval of the compound for medication purposes [16]. There are different ways to classify the toxicity of a compound. For example, based on systemic toxic effects, the common toxicity types include acute toxicity, sub-chronic toxicity, chronic toxicity, carcinogenicity developmental toxicity and genetic toxicity [22]. On the other hand, based on the toxicity effects area, toxicity can also be classified as hepatotoxicity, ototoxicity, ocular toxicity, etc. [15]. Therefore, there is a great demand for accurate, low-cost and time-saving toxicity prediction methods for different toxicity categories.

Toxicity of a chemical compound is associated with its chemical structure [17]. A good example is the chiral compounds. This kind of compounds and their isomers have highly similar structures but only slight differences in molecular geometry. Their differences cause them to possess different biological properties. For example, the drug Dopa is a compound for treating the Parkinson disease. The d-isomer form of this compound has severe toxicity whereas the l-isomer form does not [12]. Therefore, only its levorotatory form can be used for medical treatments. This property-structure relationship is often described as quantitative structure-activity relationship (QSAR) and have been widely used in the prediction of different properties of compounds [4,24]. Based on the same idea, toxicities of a compound, being one of the most concerned properties, can be predicted via computational means as a way to select more promising candidates before undertaking further biological experiments.

The Simplified Molecular Input Line Entry System, also called SMILES [20, 21], is a linear representation of a chemical compound. It is a short ASCII string describing the composition, connectivity, and charges of atoms in a compound. An example is shown in Fig. 1. The compound is called Morphine; it is originated from the opiate family and is found to exist naturally in many plants and animals. Morphine has been widely used as a medication to relief acute and chronic pain of patients. Nowadays, compounds are usually converted into their SMILES strings for the purpose of easy storage into databases or for other computational processing such as machine learning. Common molecular toolkits such as RDkit [8] and OpenBabel [13] can convert a SMILES string to its 2D and 3D structures, and vice versa.

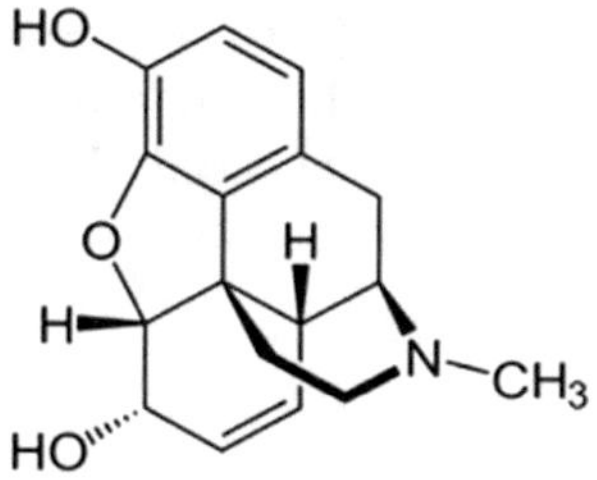

SMILES: CN1CCC23C4C1CC5=C2C(=C(C=C5)O)OC3C(C=C4)O

**Fig. 1.** The morphine structure and its SMILES representation.

In recent years, machine learning has become the mainstream technique in natural language processing (NLP). Among all machine learning applications for NLP, text classification is the most widely studied. Based on the input text sentences, a machine learning-based NLP model analyzes the organization of words and the types of words in order to categorize the given text. Two pioneering NLP methods are textCNN [6] and ConvNets [26]. The former method introduced a pretrained embedding layer to encode words of input sentences into fixed-size feature vectors with padding. Then, feature vectors of all words were combined to form a sentence matrix that was fed into a standard convolutional neural network (CNN) model. This work was considered a breakthrough at that time and accumulated over 5800 citations since 2014 (as per Google Scholar). Another spotlight paper in NLP for text classification is ConvNets [26]. Instead of analyzing words in a sentence, this model exploited simple one-hot encoding method at the character level for 70 unique characters in sentence analysis. The success of these methods in NLP shed lights to other applications that have only texts as raw data.

Compound toxicity prediction can be considered as a classification problem too. Recently, Hirohara et al. [3] proposed a new CNN model for toxicity classification based on character-level encoding. In this work, each SMILES character is encoded into a 42-dimensional feature vector. The CNN model based on this simple encoding method achieved an area-under-curve (AUC) value of 0.813 for classification of 12 endpoints using the TOX21 dataset [18]. The best AUC score in TOX21 challenge is 0.846 which is achieved by DeepTox [10]. Despite of its higher accuracy, the DeepTox model is extremely complex. It requires heavy feature engineering from a large pool of static and dynamic features derived from the compounds or indirectly via external tools. The classification model is ensemble-based combining deep neural network (DNN) with multiple layers of hidden nodes ranging from $2^{10}$ to $2^{14}$ nodes. The train dataset for this highly complex model was comprised of over 12,000 observations and superior predictive performance was demonstrated.

Besides classification, toxicity prediction can be seen as a regression problem when the compound toxicity level is of concern. Like other QSAR problems, toxicity regression is a highly challenging task due to limited data availability and noisiness of the data. With limited data, the use of simpler model architecture is preferred to avoid the model being badly overfitted. In this work, we have focused on the regression of acute oral toxicity of chemical compounds. Two recent works [5,23] were found to solve this problem where the maximally achievable $R^2$ is only 0.629 [5].

## 2    Materials and Methods

### 2.1    Dataset

In this study, we developed a regression model for acute oral toxicity prediction. The prediction task is to estimate the median lethal dose, $LD_{50}$, of the compound; this is the dose required to kill half the members of the tested population. A small $LD_{50}$ value indicates high toxicity level whereas a large $LD_{50}$ value indicates low toxicity level of the compound. Based on the $LD_{50}$ value, compounds can be categorized into four levels as defined by the United States Environmental Protection Agency (EPA) (see Table 1).

**Table 1.** Four categories of compound oral toxicity (mg/kg) defined by the United States Environmental Protection Agency.

| Category | Description | Range |
| --- | --- | --- |
| Category I | Highly toxic and severely irritating | $LD_{50} \leq 50$ |
| Category II | Moderately toxic and moderately irritating | $50 < LD_{50} \leq 500$ |
| Category III | Slightly toxic and slightly irritating | $500 < LD_{50} \leq 5000$ |
| Category IV | Practically non-toxic and not an irritant | $5000 < LD_{50}$ |

The rat acute oral toxicity dataset used in this study was kindly provided by the author of TopTox [23]. This dataset was also used in the recent study of computational toxicity prediction by Karim et al. [5]. For $LD_{50}$ prediction task, the dataset contains 7413 samples; out of which 5931 samples are for training and 1482 samples are for testing. The original train/test split was deliberately made to maintain similar distribution of the train and test datasets to facilitate learning and model validation. It is noteworthy that as the actual $LD_{50}$ values were in a wide range (train set: 0.042 mg/kg to 99947.567 mg/kg, test set: 0.020 mg/kg to 114062.725 mg/kg), the $LD_{50}$ values were first transformed to mol/kg format, and then scaled logarithmically to $-log_{10}(LD_{50})$. Finally, the processed experimental values range from 0.470 to 7.100 in the train set and 0.291 to 7.207 in the test set.

### 2.2    Binary Encoding Method for SMILES (BES)

As a SMILES string is not an understandable input format for general machine learning methods, it needs to be converted or encoded into a series of numerical values. Ideally, these values should capture the characteristics of the compound and correlates to the interested observables. The most popular way to encode a SMILE is to use molecular fingerprints such as Molecular Access System (MACCS) and extended connectivity fingerprint (ECFP). However, fingerprint algorithms generate high dimensional and sparse matrices which make learning difficult.

Here, in order to solve the regression task for oral toxicity prediction. Inspired by the work of Hirohara et al. [3], we proposed the modified Binary Encoding method for SMILES, named BES for short. In BES, each character is encoded by a binary vector of 56 bits. Among them 26 bits are for encoding the SMILES alphabets and symbols by the one-hot encoding approach; 30 bits are for encoding various atomic properties including number of bonded hydrogens, formal charge, valence, ring atom, degree, aromaticity, chirality, and hybridization. The feature types and corresponding size of the feature is listed in Table 2.

**Table 2.** The proposed binary encoding scheme called BES. Each SMILES character is encoded into a vector of 56 bits; each value, either a 1 or 0, represents the existence of that feature type and the element that it has.

| Feature type | Element | Bit(s) |
|---|---|---|
| Symbol in SMILES | ( ) | 2 |
| | [ ] | 2 |
| | | 1 |
| | : | 1 |
| | = | 1 |
| | # | 1 |
| | \ | 1 |
| | / | 1 |
| | @ | 1 |
| | + | 1 |
| | − | 1 |
| Number in SMILES | Atom charge (2–7) | 6 |
| | Ring begin (yes/no) | 1 |
| | Ring end (yes/no) | 1 |
| Atom type | C, H, O, N, others | 5 |
| Others | Surrounding hydrogen number (0–3) | 4 |
| | Atom formal charge $(-1, 0, 1)$ | 3 |
| | Valence (1–6) | 6 |
| | Ring atom (yes/no) | 1 |
| | Degree (1–5) | 5 |
| | Aromaticity (yes/no) | 1 |
| | Chirality (R/S/others) | 3 |
| | Type of hybridization | 7 |
| **Total** | - | **56** |

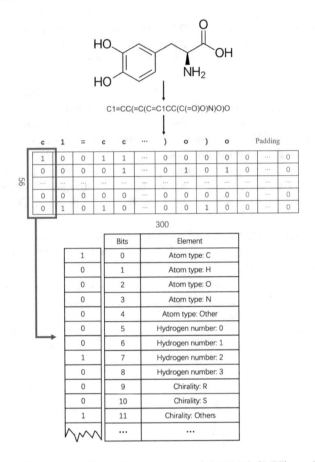

**Fig. 2.** Illustration of the Binary-Encoded SMILES (BES) method.

As the maximum length of SMILES strings in our dataset is 300, the size of the feature matrix for one SMILES string was defined to be 56 × 300. For a SMILES string that is shorter than 300 in length, zero padding was applied. Figure 2 illustrates how BES works.

## 2.3    Model Architecture

Our prediction model is a conventional CNN model with convolutional layers to extract features, pooling layers to reduce dimensionality of the feature matrix and to prevent overfitting, and a multi-layer neural network to correlate features to $LD_{50}$ values. To decide the model architecture and to tune hyperparameters of the model, a grid search method was employed. Table 3 shows the hyperparameters and their ranges of values within which the model was optimized. In each grid search process, the model training was run for 500 epochs and the mean-squared error (MSE) loss of the model in 5-fold cross validation was used

**Table 3.** Tuning options and optimal values for hyperparameters.

| Hyperparameter | Candidate value | Optimal value |
|---|---|---|
| Number of filter (Conv 1) | 256–1024 | 512 |
| Number of filter (Conv 2) | 256–1024 | 1024 |
| Activation function (Conv) | ReLU, Sigmoid | ReLU |
| Activation function (FC) | ReLU, Sigmoid | ReLU |
| Batch size | 16, 32, 64, 128 | 32 |
| Batch normalization (BN) | Yes, No | Yes |
| Dropout | Yes, No | No |
| Optimizer | Adam, SGD | Adam |
| Learning rate | 0.1, 0.01, 0.001 | 0.01 |
| Max epoch | 500,1000,1500 | 1000 |

as a criteria for model selection. The optimal parameters are also presented in Table 3. The final production model was trained using the optimal parameters and the entire train dataset. The maximum training epoch was 1000; early stop method was used to prevent the problem of overfitting.

The architecture of our optimized CNN model is presented in Fig. 3. The model contains two convolutional layers (Conv) with 512 and 1024 filters respectively. After each convolutional layer is an average pooling layer and a batch normalization layer (BN). Then, a max pooling layer is used before the learned features fed into the fully connected layers (FC). Four FCs containing 2048, 1024, 512, and 256 hidden nodes were found to be the optimal combination for toxicity prediction and the ReLU function is used to generate the prediction output.

## 2.4   Implementation

All implementations were done using Python 3.6.9 with the following libraries: Anaconda 4.7.0, RDKit v2019.09.2.0, Pytorch 1.2.0 and CUDA 10.0. We used `GetTotalNumHs`, `GetFormalCharge`, `GetChiralTag`, `GetTotalDegree`, `IsInRing`, `GetIsAromatic`, `GetTotalValence` and `GetHybridization` functions from RDkit to calculate atom properties. Our model was trained and tested in a workstation equipped with two NVIDIA Tesla P100 GPUs.

# 3   Results

Training of the final production model was performed using the optimal parameters obtained from the result of our extensive grid search. Figure 4 shows the evolution of MSE over the number of training cycles. The training stopped at the 900-th epoch with MSE of 0.016. Table 4 shows the performances of our model

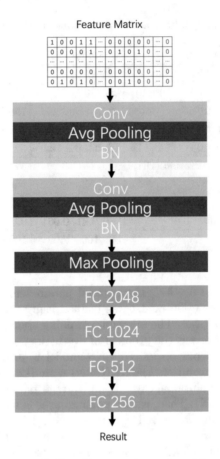

**Fig. 3.** Proposed CNN architecture for oral toxicity prediction.

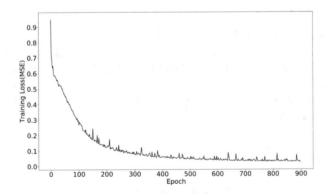

**Fig. 4.** MSE loss of training BESTox model with optimal hyperparameters in Table 3.

**Table 4.** Predictive performances of BESTox in the train and test datasets. $R^2$ is the squared Pearson correlation coefficient, RMSE is root-mean-squared error and MAE is mean absolute error.

| Performance | Dataset | Size | $R^2$ | RMSE | MAE |
|---|---|---|---|---|---|
| Training | Train set | 5931 | 0.982 | 0.126 | 0.084 |
| Testing | Test set | 1482 | 0.619 | 0.603 | 0.433 |

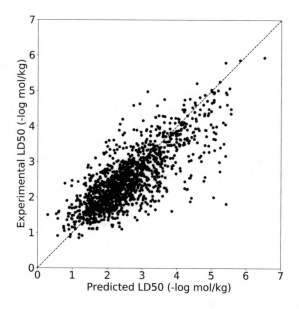

**Fig. 5.** Scatterplot of BESTox prediction on 1482 test data ($R^2 = 0.619$).

in the train and test sets. The training performance is excellent which gives $R^2$ of 0.982 as all the data was used to construct the model. For the test set, the model predicts with $R^2$ of 0.619, RMSE of 0.603, and MAE of 0.433.

Figure 5 shows the scatterplot of BESTox prediction on the test data. We can see that prediction is better for compounds with lower toxicity (lower $-log_{10}(LD_{50})$) and worse for those with higher toxicity. This may be due to fewer data available in the train set for higher toxicity compounds. Thus, we also tested our model on samples with target values less than 3.5 in the test set (1255 samples out of total 1482 samples, the sample coverage is more than 84%). In this case, the performance of our model is improved: RMSE is decreased from 0.603 to 0.516 and MAE is reduced from 0.433 to 0.385.

**Table 5.** Performance comparison of our model to two existing acute oral toxicity prediction methods: TopTox [23] and DT+SNN [5]. Performance data of these methods were obtained from the original literature.

| Model | $R^2$ | RMSE | MAE | Ref. |
|---|---|---|---|---|
| ST-DNN (TopTox) | 0.614 | 0.601 | 0.436 | [23] |
| DT+SNN | 0.629 | – | – | [5] |
| BESTox | 0.619 | 0.603 | 0.433 | This study |

Table 5 presents the comparative performance of BESTox to two existing acute oral toxicity prediction models, the ST-DNN model from TopTox and the DT+SNN model from Karim et al. [5]. Results show that our model is slightly better than ST-DNN with respect to $R^2$ and MAE. The best performed model is DT+SNN which has a correlation of 0.629; but RMSE and MAE were not provided in the original study.

The closeness of the performance metrics of BESTox to two existing models suggest that our model performs on par with them. Nevertheless, it should be mentioned that while our model has employed simple features and relatively simple model architecture, ST-DNN and DT+SNN relied on highly engineered input features and complex ensemble-based model architectures. For ST-DNN [23], they combined 700 element specific topological descriptors (ESTD) and 330 auxiliary descriptors as candidates to generate the feature vectors for prediction (our model uses only 56 features). In addition, their model included ensemble of two different types of classifiers, namely, deep neural network (DNN) and gradient boosted decision tree (GBDT). Combining predictions from several classifiers is an easy way to improve prediction accuracy, however, the complexity introduced into the model makes the already "black box model" more difficult to understand. For the recent DT+SNN model [5], they used decision trees (DT) to select 817 different descriptors generated from the PaDEL tools [25]. Although their shallow neural network (SNN) architecture required short model training time, more time was spent on feature generation and selection. Different combination of features were used depending on the tasks to be predicted, which had high computational cost. Here, BESTox has achieved results comparable to these more complex models with simple binary features and model architecture, showing the power of our method.

## 4    Conclusion

In this paper, we present our new method BESTox for acute oral toxicity prediction. Inspired by NLP techniques for text classification, we have designed a simple character-level encoding method for SMILES called the binary-encoded SMILES (BES). We have developed a shallow CNN to learn the BES matrices to predict the $LD_{50}$ values of compounds. We trained our model on the rat acute oral toxicity data, tested and compared to two other existing models. Despite the simplicity of our method, BESTox has achieved a good performance with $R^2$ of 0.619, comparable to the single-task model proposed by TopTox [23] but slightly inferior to the hybrid decision tree and shallow neural network model by Karim et al. [5].

Future improvement of BESTox will be focused on extending the scope of datasets. As shown in the work of Wu et al. [23], multitask learning can improve performance of prediction models due to availability of more data on different toxicity effects. The idea of multitask technique is to train a model with multiple training sets; each set corresponds to one toxicity prediction task. Feeding the learners with different toxicity data helps them to learn common latent features of molecules offered by different datasets.

**Acknowledgments.** This work was supported by University of Macau (Grant no. MYRG2017-00146-FST).

# References

1. Bailey, J., Balls, M.: Recent efforts to elucidate the scientific validity of animal-based drug tests by the pharmaceutical industry, pro-testing lobby groups, and animal welfare organisations. BMC Med. Ethics **20**, 16 (2019)

2. Dean, A., Lewis, S.: Screening: Methods for Experimentation in Industry, Drug Discovery, and Genetics. Springer, Cham (2006). https://doi.org/10.1007/0-387-28014-6

3. Hirohara, M., Saito, Y., Koda, Y., Sato, K., Sakakibara, Y.: Convolutional neural network based on SMILES representation of compounds for detecting chemical motif. BMC Bioinform. **19**, 526 (2018)

4. Idakwo, G., et al.: A review on machine learning methods for in silico toxicity prediction. J. Environ. Sci. Health Part C **36**(4), 169–191 (2018)

5. Karim, A., Mishra, A., Newton, M.H., Sattar, A.: Efficient toxicity prediction via simple features using shallow neural networks and decision trees. ACS Omega **4**(1), 1874–1888 (2019)

6. Kim, Y.: Convolutional neural networks for sentence classification. arXiv preprint arXiv:1408.5882 (2014)

7. Kubinyi, H., Mannhold, R., Timmerman, H.: Virtual Screening for Bioactive Molecules, vol. 10. Wiley, Hoboken (2008)

8. Landrum, G., et al.: RDkit: open-source cheminformatics (2006)

9. Llanos, E.J., Leal, W., Luu, D.H., Jost, J., Stadler, P.F., Restrepo, G.: Exploration of the chemical space and its three historical regimes. Proc. Natl. Acad. Sci. **116**(26), 12660–12665 (2019)

10. Mayr, A., Klambauer, G., Unterthiner, T., Hochreiter, S.: DeepTox: toxicity prediction using deep learning. Front. Environ. Sci. **3**, 80 (2016)

11. McInnes, C.: Virtual screening strategies in drug discovery. Curr. Opin. Chem. Biol. **11**(5), 494–502 (2007)

12. Nguyen, L.A., He, H., Pham-Huy, C.: Chiral drugs: an overview. Int. J. Biomed. Sci. IJBS **2**(2), 85 (2006)

13. O'Boyle, N.M., Banck, M., James, C.A., Morley, C., Vandermeersch, T., Hutchison, G.R.: Open Babel: an open chemical toolbox. J. Cheminform. **3**(1), 33 (2011)

14. Oprea, T.I., Matter, H.: Integrating virtual screening in lead discovery. Curr. Opin. Chem. Biol. **8**(4), 349–358 (2004)
15. Quintanilha, J.C.F., Berlofa, M.: New promising approaches to treatment of chemotherapy-induced toxicities. AvidScience Chemother. 2–52 (2017)
16. Raies, A.B., Bajic, V.B.: In silico toxicology: computational methods for the prediction of chemical toxicity. Wiley Interdiscip. Rev. Comput. Mol. Sci. **6**(2), 147–172 (2016)
17. Roy, K., Kar, S., Das, R.: Chapter 7–validation of QSAR models. Understanding the basics of QSAR for applications in pharmaceutical sciences and risk assessment, pp. 231–289 (2015)
18. Tice, R.R., Austin, C.P., Kavlock, R.J., Bucher, J.R.: Improving the human hazard characterization of chemicals: a TOX21 update. Environ. Health Perspect. **121**(7), 756–765 (2013)
19. Ting, N.: Dose Finding in Drug Development. Springer, Cham (2006). https://doi.org/10.1007/0-387-33706-7
20. Weininger, D.: SMILES, a chemical language and information system. 1. Introduction to methodology and encoding rules. J. Chem. Inf. Comput. Sci. **28**(1), 31–36 (1988)
21. Weininger, D., Weininger, A., Weininger, J.L.: SMILES. 2. Algorithm for generation of unique SMILES notation. J. Chem. Inf. Comput. Sci. **29**(2), 97–101 (1989)
22. Wexler, P., Gad, S.C., et al.: Encyclopedia of Toxicology. Academic Press, Cambridge (1998)
23. Wu, K., Wei, G.W.: Quantitative toxicity prediction using topology based multi-task deep neural networks. J. Chem. Inf. Model. **58**(2), 520–531 (2018)
24. Wu, Y., Wang, G.: Machine learning based toxicity prediction: from chemical structural description to transcriptome analysis. Int. J. Mol. Sci. **19**(8), 2358 (2018)
25. Yap, C.W.: Padel-descriptor: an open source software to calculate molecular descriptors and fingerprints. J. Comput. Chem. **32**(7), 1466–1474 (2011)
26. Zhang, X., Zhao, J., LeCun, Y.: Character-level convolutional networks for text classification. In: Advances in Neural Information Processing Systems, pp. 649–657 (2015)

# Stratified Test Alleviates Batch Effects in Single-Cell Data

Shaoheng Liang[1,2], Qingnan Liang[3], Rui Chen[3], and Ken Chen[2(✉)]

[1] Rice University, Houston, TX 77005, USA
[2] The University of Texas MD Anderson Cancer Center,
Houston, TX 77030, USA
{sliang3,kchen3}@mdanderson.org
[3] Baylor College of Medicine, Houston, TX 77030, USA

**Abstract.** Analyzing single-cell sequencing data across batches is challenging. We find that the Van Elteren test, a stratified version of Wilcoxon rank-sum test, elegantly mitigates the problem. We also modified the common language effect size to supplement this test, further improving its utility. On both simulated and real patient data we show the ability of Van Elteren test to control for false positives and false negatives. The effect size also estimates the differences between cell types more accurately.

**Keywords:** scRNA-seq analysis · Differential expression analysis · Batch effect · Wilcoxon rank-sum test · Van Elteren test

## 1 Introduction

Large-scale studies such as the Human Cell Atlas [15] involve hundreds of laboratories, thousands of patients, and millions of cells, bringing about opportunities and challenges in analyses. When comparing cell types or groups, discrepancies across experiments and differences among participants lead to omissions and false discoveries in differentially expressed genes. Even the trend (upregulated or downregulated) can be reversed in a phenomenon called Simpson's paradox [1]. Although multiple methods have been proposed to tackle such batch effects, no such option for the widely used Wilcoxon rank-sum test [11,18] has been applied to single-cell studies, to the best of our knowledge. Here, we show that the stratified rank-sum test (known as Van Elteren test [17]) and our modified common language effect size may benefit single-cell studies.

We briefly review and conceptually compare related works on correcting batch effect in Sect. 1.1. Then, in Sect. 2, we revisit Wilcoxon rank-sum test,

Supported by grant number 2018-182735 to KC, Human Breast Cell Atlas Seed Network Grant (HCA3-0000000147) to NN and KC from the Chan Zuckerberg Initiative DAF, an advised fund of Silicon Valley Community Foundation, grant RP180248 to KC from Cancer Prevention & Research Institute of Texas, and the Cancer Center Support Grant P30 CA016672 to PP from the National Cancer Institute.

© Springer Nature Switzerland AG 2020
C. Martín-Vide et al. (Eds.): AlCoB 2020, LNBI 12099, pp. 167–177, 2020.
https://doi.org/10.1007/978-3-030-42266-0_13

and introduce the Van Elteren test supplemented by our direct extension of the common language effect size [5,12]. Simulation studies and applications to real data in Sect. 3 show that the test controls for the batch effects and leads to more accurate biological discovers, compared with Wilcoxon rank-sum test. More discussions and explanations, are shown in Sect. 4.

## 1.1    Related Works

Mainstream methods to mitigate batch effect fall into two categories, batch correction methods and batch-aware statistical tests. The former includes methods reducing batch effect in the data to facilitate downstream analysis, while the latter includes analyses that control for the batch effect.

**Batch Correction Methods.** Batch correction methods eliminate the discrepancy among batches to create an integrated dataset. The most conspicuous manifestation of batch effect is splitting one cell type into multiple clusters. To solve this problem, many methods match and combine clusters across samples based on similarities. A commonly adopted one, Mutual nearest neighbor (MNN) [3], uses similar cells across datasets as anchors, and based on them correct the gene expression of other cells. Scanorama [4] and Seurat [16] are both based on the MNN methodology. Another method, Harmony [7], iteratively corrects the data by clustering the cells and moving neighboring clusters toward each other. These methods typically produce a unified data matrix, which can be conveniently used in visualization and downstream analysis. However, these empirical corrections usually lack negative control and raise uncertainty in the discovery [14].

**Batch-Aware Statistical Tests.** Instead of manipulating the data directly, statistical methods may handle batch effect by considering it as a covariate in the model. This is possible in Student's t-test, Poisson test, negative binomial test, etc. Notably, all these tests are parametric, meaning that a distribution must be given in advance. However, the debate of the true distribution of single-cell gene expression has never ceased, which is a reason why the nonparametric Wilcoxon rank-sum test is widely used. To allow modeling covariates, one may use a generalized version of rank-sum test, the proportional odds model [2]. However, modeling batches by using a covariate also makes unnecessary assumptions upon them. Stratification, which only combines statistics from batches, is the "as simple as possible, but no simpler" way to handle batches. The Van Elteren test we use, is the stratified version of Wilcoxon rank-sum test.

It is worth noting that methods like scVI [9] have combined statistical modeling with batch effect correction. However, the effect of batches is modeled by a black-box neural network, making the interpretation elusive.

# 2 Methods

## 2.1 Wilcoxon Rank-Sum Test

We briefly revisit the Wilcoxon rank-sum test (also known as Mann–Whitney U test) [11,18]. The test statistics $U$ is defined as

$$U = \sum_{i=1}^{n_1} \sum_{j=1}^{n_2} \left( \mathbb{1}_{a_i > b_j} + \frac{1}{2} \mathbb{1}_{a_i = b_j} \right), \tag{1}$$

where $A = \{a_i\}_{i=1}^{n_1}$ and $B = \{b_j\}_{j=1}^{n_2}$ are the two samples to be compared (e.g., two cell types in one experiment), with sample sizes $n_1$ and $n_2$, respectively. Function "$\mathbb{1}$" takes value 1 when its condition holds true, and 0 otherwise. When $n_1$ and $n_2$ are both at least 10, which is common in single-cell studies, the distribution of $U$ approximately follows a normal distribution $\mathcal{N}(\mu, \sigma^2)$ where

$$\mu = \frac{n_1 n_2}{2} \tag{2}$$

and

$$\sigma^2 = \frac{n_1 n_2}{12} \left( (n+1) - \sum_{i=1}^{k} \frac{t_i^3 - t_i}{n(n-1)} \right), \tag{3}$$

in which $t_i$ stands for ties (corresponding to the second term in Eq. 1).

## 2.2 Van Elteren Test

The Van Elteren test [17] is the stratified version of Wilcoxon rank-sum test. For example, if there are $m$ patients, they maybe treated as strata. In that case, a $U$ statistic may be obtained from each patient $g \in \{1, \cdots, m\}$, denoted as $U_g \sim \mathcal{N}(\mu_g, \sigma_g^2)$. A new statistics $V$ is constructed by

$$V = \frac{\left[ \sum_{g=1}^{m} w_g (U_g - \mu_g) \right]^2}{\sum_{g=1}^{m} w_g^2 \sigma_g^2} \sim \chi_1^2, \tag{4}$$

where $w_g$ is a weight for each sample to be discussed later. When $m = 1$, the formula degenerates to $V = (U_g - \mu_g)^2 / \sigma_g^2 \sim \chi_1^2$, which is consistent with the rank-sum test.

**Weights.** As discussed by Van Elteren [17], the weights $w_g$ can be assigned in different ways. It should be noted that the $U_g$ for a batch $g$ ranges from 0 to $n_{g1} n_{g2}$, the product of two sample sizes in the batch. Should the weights all be equal, a patient with more cells available will dominate the test results. It is proven in [17] that weight

$$w_g = \frac{1}{n_{g1} n_{g2}} \tag{5}$$

eliminates such effect, and a test utilizing such weight is thus named as "design-free test". However, given that a batch with more instances available (e.g., a patient with more cells sequenced) may be more convincing, another weight

$$w_g = \frac{1}{n_{g1} + n_{g2} + 1} \tag{6}$$

is introduced, which gives more power to larger batches. It also effectively assigns larger weights to batches whose samples are more balanced, when the batch sizes are the same. It is shown in [17] that this choice yields largest statistical power against randomized alternatives, and is thus named as "locally-best test". The comparison of two weights are shown in Fig. 1.

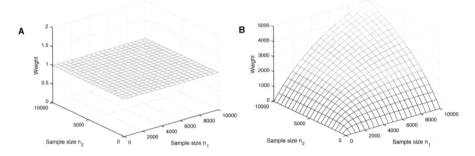

**Fig. 1.** The weight for each batch when using the (A) design-free test and (B) locally-best test. For design-free test all the batches have equal weights, while for locally-best test higher weights are given to batches with higher and balanced sample sizes.

**Effect Sizes.** For Wilcoxon rank-sum test, a simple definition of effect size is

$$f = \frac{U}{n_1 n_2}, \tag{7}$$

which is centered at 50%, meaning the probability $P(a > b)$ when $a$ and $b$ are randomly drawn from sample $A$ and sample $B$, respectively. An effect size greater than 50% generally means that $A$ is higher, and vice versa. It may be easily extended for Van Elteren test by taking average using desired weights. For the design-free test, the effect size is

$$f = \frac{1}{m} \sum_{g=1}^{m} \frac{U_g}{n_{g1} n_{g2}}, \tag{8}$$

as all batches are treated equally regardless of the sample sizes. It may be interpreted as the probability of $P(a_g > b_g)$ for $a_g$ and $b_g$ randomly drawn from

$A_g$ and $B_g$, after randomly choosing a batch $g$. For the locally-best test, the effect size becomes

$$f = \frac{1}{\sum_{g=1}^{m} \frac{n_{g1} n_{g2}}{n_{g1} + n_{g2} + 1}} \sum_{g=1}^{m} \frac{n_{g1} n_{g2}}{n_{g1} + n_{g2} + 1} \frac{U_g}{n_{g1} n_{g2}} \tag{9}$$

$$= \frac{\sum_{g=1}^{m} \frac{U_g}{n_{g1} + n_{g2} + 1}}{\sum_{g=1}^{m} \frac{n_{g1} n_{g2}}{n_{g1} + n_{g2} + 1}}, \tag{10}$$

which changes the probability of choosing a group $g$ to be in proportion to $\frac{n_{g1} n_{g2}}{n_{g1} + n_{g2} + 1}$, giving higher weights to batches with higher and balanced sample sizes (Fig. 1). Generally, any $w_g$ may be used to define $f$, as in

$$f = \frac{\sum_{g=1}^{m} U_g w_g}{\sum_{g=1}^{m} n_{g1} n_{g2} w_g}, \tag{11}$$

the two previous options being its special cases.

## 3   Results

We implemented the Van Elteren test with the effect size in R, available at our GitHub repository (https://github.com/KChen-lab/stratified-tests-for-seurat), based on Seurat 3.0 by utilizing its differential expression analysis part (but irrelevant to the data integration) [16]. When there are two groups of cells, A and B, denoted as type, and patient identity, denoted as batch, the Van Elteren Test can be called as follows.

```
FindMarkers(obj, ident.1 = 'A', ident.2 = 'B', group.by = 'type',
            test.use = 'VE', logfc.threshold = 0,
            latent.vars = 'batch', genre = 'locally-best')
```

The genre may be set to either locally-best or design-free, as introduced in Sect. 2, based on which p-values and effect sizes are calculated. Typical results are shown in Table 2. An effect size of larger than 0.5 indicates a higher expression in cell type A, and vice versa. The avg_logFC, average logarithmic fold changes, are calculated automatically by Seurat, where a positive value indicates a higher expression. It may show different trends compared with the effect sizes. Generally, the effect sizes are more indicative after controlling for the batch effect.

### 3.1   Simulation Study

We simulated datasets to illustrate the key utilities of Van Elteren test. The parameters are specified in Table 1. Poisson distribution is used to model sequencing depth. Visualization is available in Fig. 2 for illustration. We assume that the library size of each sample is equalized by other genes beyond the simulated ones.

The testing result of Van Elteren test and Wilcoxon rank-sum test is shown in Table 2. Trend (A over B) are indicated by arrows and insignificant p-values are grayed out. For Van Elteren test, the locally-best version and the design-free version return very similar results.

**Table 1.** Simulated datasets

| Patient | Number 1 | | Number 2 | |
|---|---|---|---|---|
| Cell type | A | B | A | B |
| Cell amount | 101 | 30 | 31 | 100 |
| Gene1 | $\mathrm{Pois}(\lambda = 9)$ | $\mathrm{Pois}(\lambda = 10)$ | $\mathrm{Pois}(\lambda = 8)$ | $\mathrm{Pois}(\lambda = 9)$ |
| Gene2 | $\mathrm{Pois}(\lambda = 8)$ | $\mathrm{Pois}(\lambda = 6)$ | $\mathrm{Pois}(\lambda = 15)$ | $\mathrm{Pois}(\lambda = 14)$ |
| Gene3 | $\mathrm{Pois}(\lambda = 3)$ | $\mathrm{Pois}(\lambda = 3)$ | $\mathrm{Pois}(\lambda = 5)$ | $\mathrm{Pois}(\lambda = 5)$ |
| Gene4 | $\mathrm{Pois}(\lambda = 5)$ | $\mathrm{Pois}(\lambda = 6)$ | $\mathrm{Pois}(\lambda = 5)$ | $\mathrm{Pois}(\lambda = 6)$ |

**Suppressing False Negatives.** Batch effect may introduce false negatives, where a significantly differentially expressed gene is overshadowed. For gene 1, on which cell type B always have higher expression on both patients, the Wilcoxon rank-sum test did not pass the threshold of 0.05 to reject the null hypothesis, while Van Elteren test yields a significant p-value. The effect size, smaller than 0.5, also correctly suggests that the expression of gene 1 in cell type B is higher than that in cell type A, compared with the average logarithmic fold change, which wrongfully indicates otherwise.

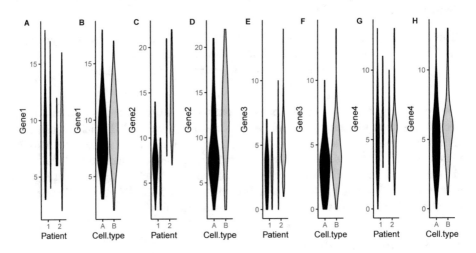

**Fig. 2.** Illustration of four simulated genes. Two shades correspond to two cell types (dark: cell type A; light: cell type B). For each gene, the left panel is stratified by patients and the right panel shows aggregated distribution.

**Table 2.** Results of the two tests in simulation studies

| | P-value | | | Effect size | | |
|---|---|---|---|---|---|---|
| | Wilcoxon | Van Elteren | | Log fold change | Van Elteren $f$ | |
| | | locally-best | design-free | | locally-best | design-free |
| Gene1 | 6.336E−02 | 3.119E−03 | 3.109E−03 | 0.673 ↑ | 0.375 ↓ | 0.375 ↓ |
| Gene2 | 3.700E−08 | 8.469E−03 | 8.193E−03 | −1.182 ↓ | 0.611 ↑ | 0.611 ↑ |
| Gene3 | 5.770E−05 | 7.831E−01 | 7.905E−01 | −3.664 ↓ | 0.488 ↓ | 0.489 ↓ |
| Gene4 | 2.465E−03 | 2.245E−03 | 2.416E−03 | −0.817 ↓ | 0.373 ↓ | 0.372 ↓ |

**Suppressing Reversed Conclusions.** Batch effect may also lead to reversed conclusion (i.e., which cell type has higher expression). For gene 2, on which cell type A always have higher expression value, both tests reject the null hypothesis. However, the effect size of Van Elteren test, larger than 0.5, correctly identifies that the expression of gene 2 in cell type A is higher than that in cell type B, while the average logarithmic fold change wrongfully indicates otherwise.

**Suppressing False Positives.** False discoveries are also possible outcome of batch effect. As is shown in gene 3, the distribution of both cell types are exactly the same in each patient. Nevertheless, Wilcoxon test yields a very significant p-value. The average logarithmic fold change also has a large magnitude. Van Elteren test returns a p-value of 0.7831, together with a effect size close to 0.5, suggesting that the difference is neither significant nor large.

**Consistency.** As a negative control, when the three issues above are not present, p-value from Van Elteren test is consistent with Wilcoxon rank-sum test, as is shown by gene 4. The effect size and the log fold change also both show that the cell type B has higher expression in gene 4.

## 3.2 Retina Data

We have tested the Van Elteren test on real retina single-cell gene expression data gathered from three patients [8]. Two regions, macula (i.e., the center area) and peripheral, are labeled in the data. We question which genes differentially express for the same cell type between two regions. We run Wilcoxon rank-sum test and Van Elteren test on 2,295 rod cells and 203 cone cells. We compare the results in Fig. 3, where genes with large differences in p-values between two tests are labeled with gene names.

**Rod Cells.** The results of two tests are largely comparable, showing a diagonal pattern. Meanwhile, some exceptions are present (see Table 3 for the p-values

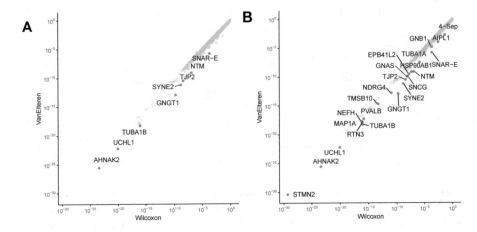

**Fig. 3.** Comparison of p-values returned from Wilcoxon rank-sum test and Van Elteren test on (A) rod cells and (B) cone cells. Each dot is a gene, whose p-value from Wilcoxon rank-sum test and Van Elteren test are shown by its x-coordinate and y-coordinate, respectively. Genes with largely changed p-values ($10^2$ for rod cells and $10^1$ for cone cells) are labeled with gene names.

and effect sizes), among which we observed that p-values for gene *GNGT1* and *SYNE2* change the most.

For *GNGT1*, the reversed conclusion effect is also observed, as the Van Elteren $f$ effect size suggests that the peripheral region has a higher expression, while the log fold change indicates otherwise. We further inspected the distributions to validate and interpret the differences. In Fig. 4A, the left panel does show generally higher expression of *GNGT1* in each individual patient, while the aggregated distribution on the right shows a reversed effect, which is an instance of the aforementioned Simpson's paradox. For *SYNE2* (Fig. 4B), conspicuous discrepancy among batches is also shown, which leads to a less precise rank-sum test result. Indeed, these two genes were found playing roles in macular degeneration diseases [6, 10].

**Cone Cells.** For cone cells, we observed similar results. Some genes show changes while most genes are consistent across the tests. The p-values and effect sizes are shown in Table 4.

For gene *PCBP4*, Van Elteren test shows more significant p-value, and an effect size indicating smaller expression in macula, which is different from the log fold change. Decrease in *PCBP4* has also been linked with age-related macular degeneration [13]. Figure 5 shows that batch effect in distribution of *PCBP4* may have misled the rank-sum test and the logarithmic fold change.

**Table 3.** Gene with large p-value change in rod cells

|  | P-value | | | Effect size | |
|---|---|---|---|---|---|
|  | Wilcoxon | Van Elteren | Log10 ratio | Log fold change | Van Elteren $f$ |
| *GNGT1* | 1.694E−10 | 1.229E−13 | 3.14 | 89.679 ↑ | 0.390 ↓ |
| *SYNE2* | 1.451E−09 | 6.654E−12 | 2.34 | −48.321 ↓ | 0.400 ↓ |
| *TUBA1B* | 9.317E−17 | 5.521E−19 | 2.23 | 233.992 ↑ | 0.631 ↑ |
| *UCHL1* | 9.044E−21 | 6.300E−23 | 2.16 | 243.679 ↑ | 0.639 ↑ |
| *NTM* | 1.116E−07 | 8.324E−10 | 2.13 | 3.679 ↑ | 0.410 ↓ |
| *AHNAK2* | 4.179E−24 | 3.140E−26 | 2.12 | 178.679 ↑ | 0.644 ↑ |
| *SNAR-E* | 1.991E−04 | 1.664E−06 | 2.08 | Inf ↑ | 0.431 ↓ |
| *TJP2* | 3.650E−09 | 3.358E−11 | 2.04 | 75.678 ↑ | 0.411 ↓ |

**Table 4.** Gene with large p-value change in cone cells

|  | P-value | | | Effect size | |
|---|---|---|---|---|---|
|  | Wilcoxon | Van Elteren | Log10 ratio | Log fold change | Van Elteren $f$ |
| *PCBP4* | 1.734E−02 | 1.285E−03 | 1.13 | 5.397 ↑ | 0.335 ↓ |

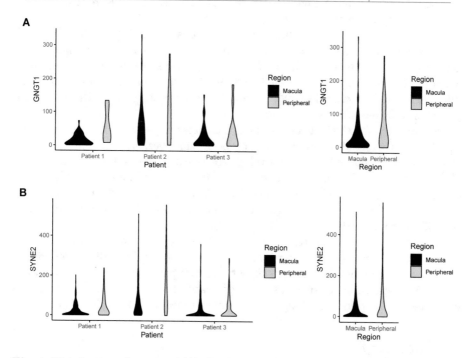

**Fig. 4.** Distribution of counts of (A) *GNGT1* and (B) *SYNE2* in rod cells. Left panels are stratified by patients and right panels show aggregated distributions.

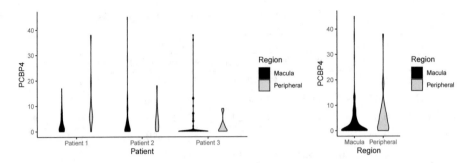

**Fig. 5.** Distribution of counts of *PCBP4* in rod cells. The left panel is stratified by patients and the right panel shows aggregated distributions.

## 4  Discussion

The results have clearly shown that Van Elteren test benefits biological studies in precisely identifying differentially expressed genes. Although the results we show do not include multiple comparison correction, Seurat 3.0 will automatically give corrected p-value based on the raw p-value using Bonferroni correction. Generally, any correction based on p-values will also apply.

The result also indicates that stratified test is a neat way to handle batch effect. Although covariate has the ability to control for explanatory variables, it is generally more suitable for continuous variables. It also casts more assumptions when modeling covariate. Stratified test, on the other hand, does not infer the influence of the discrete batches. Rather, it directly aggregates the statistical power of multiple samples.

Admittedly, for the rod cells, although changes in p-values are observed, the significance threshold was well passed by both. However, it should be noted that the retina data are collected from relatively healthy tissues and are considered clean, while Van Elteren test is expected to make a more meaningful difference on noisy pathological and tumor data. In addition, rod cell is the most populous cell in retina. For rare cell types that take smaller proportions, like the cone cells, the difference Van Elteren test makes can be crucial.

The caveat of stratified test is that for it to work the strata shall not overlap with the variable of interest. For instance, it may not find the difference, meanwhile also control for the batch effect, between two patients. Nonetheless, neither is covariate applicable to such cases. As the batch effect and biological effect are convolved, more prior knowledge is generally needed to distinguish them.

## 5  Summary

We have adopted Van Elteren test, an underappreciated statistical test, and our weighted common language effect size to single-cell sequencing data. When batch effect is severe, the test control for false positives. Otherwise, it is consistent with Wilcoxon rank-sum test. The modified common language effect size also faithfully depicts the trends. This work may increase the precision of differential expression analysis to help identify genes of interests.

# References

1. Blyth, C.R.: On Simpson's paradox and the sure-thing principle. J. Am. Stat. Assoc. **67**(338), 364–366 (1972)
2. Everitt, B., Skrondal, A.: The Cambridge Dictionary of Statistics. BusinessPro collection, 4th edn. Cambridge University Press, Cambridge (2010)
3. Haghverdi, L., Lun, A.T., Morgan, M.D., Marioni, J.C.: Batch effects in single-cell RNA-sequencing data are corrected by matching mutual nearest neighbors. Nat. Biotechnol. **36**(5), 421 (2018)
4. Hie, B., Bryson, B., Berger, B.: Efficient integration of heterogeneous single-cell transcriptomes using Scanorama. Nat. Biotechnol. **37**(6), 685 (2019)
5. Kerby, D.S.: The simple difference formula: an approach to teaching nonparametric correlation. Compr. Psychol. **3**, 1–10 (2014). 11-T
6. Kolesnikov, A.V., et al.: G-protein $\beta\gamma$-complex is crucial for efficient signal amplification in vision. J. Neurosci. **31**(22), 8067–8077 (2011)
7. Korsunsky, I., et al.: Fast, sensitive and accurate integration of single-cell data with Harmony. Nat. Methods **16**, 1–8 (2019)
8. Liang, Q., et al.: Single-nuclei RNA-seq on human retinal tissue provides improved transcriptome profiling. Nat. Commun. **10**(1), 1–12 (2019)
9. Lopez, R., Regier, J., Cole, M.B., Jordan, M.I., Yosef, N.: Deep generative modeling for single-cell transcriptomics. Nat. Methods **15**(12), 1053 (2018)
10. Maddox, D.M., et al.: A mutation in Syne2 causes early retinal defects in photoreceptors, secondary neurons, and Müller glia. Invest. Ophthalmol. Vis. Sci. **56**(6), 3776–3787 (2015)
11. Mann, H.B., Whitney, D.R.: On a test of whether one of two random variables is stochastically larger than the other. Ann. Math. Stat. **18**, 50–60 (1947)
12. McGraw, K.O., Wong, S.: A common language effect size statistic. Psychol. Bull. **111**(2), 361 (1992)
13. Meyer, J.G., Garcia, T.Y., Schilling, B., Gibson, B.W., Lamba, D.A.: Proteome and secretome dynamics of human retinal pigment epithelium in response to reactive oxygen species. Sci. Rep. **9**(1), 1–12 (2019)
14. Nygaard, V., Rødland, E.A., Hovig, E.: Methods that remove batch effects while retaining group differences may lead to exaggerated confidence in downstream analyses. Biostatistics **17**(1), 29–39 (2016)
15. Regev, A., et al.: Science forum: the human cell atlas. Elife **6**, e27041 (2017)
16. Stuart, T., et al.: Comprehensive integration of single-cell data. Cell **177**, 1888–1902 (2019)
17. Van Elteren, P.: On the combination of independent two-sample tests of Wilcoxon. Bull. Inst. Int. Staist. **37**, 351–361 (1960)
18. Wilcoxon, F.: Individual comparisons by ranking methods. In: Kotz, S., Johnson, N.L. (eds.) Breakthroughs in Statistics. Springer Series in Statistics (Perspectives in Statistics), pp. 196–202. Springer, New York (1992). https://doi.org/10.1007/978-1-4612-4380-9_16

# A Topological Data Analysis Approach on Predicting Phenotypes from Gene Expression Data

Sayan Mandal[1] , Aldo Guzmán-Sáenz[2] , Niina Haiminen[2] ,
Saugata Basu[3] , and Laxmi Parida[2](✉) 

[1] The Ohio State University, Columbus, OH, USA
[2] IBM Research, T. J. Watson Research Center, Yorktown Heights, NY, USA
parida@us.ibm.com
[3] Purdue University, West Lafayette, IN, USA

**Abstract.** The goal of this study was to investigate if gene expression measured from RNA sequencing contains enough signal to separate healthy and afflicted individuals in the context of phenotype prediction. We observed that standard machine learning methods alone performed somewhat poorly on the disease phenotype prediction task; therefore we devised an approach augmenting machine learning with topological data analysis.

We describe a framework for predicting phenotype values by utilizing gene expression data transformed into sample-specific topological signatures by employing feature subsampling and persistent homology. The topological data analysis approach developed in this work yielded improved results on Parkinson's disease phenotype prediction when measured against standard machine learning methods.

This study confirms that gene expression can be a useful indicator of the presence or absence of a condition, and the subtle signal contained in this high dimensional data reveals itself when considering the intricate topological connections between expressed genes.

**Keywords:** Topological data analysis · Gene expression · Phenotype prediction · Parkinson's disease

## 1   Introduction

Our aim in this work was to investigate if the expression of protein-coding genes provide sufficient information to classify an individual with respect to a phenotype, in particular Parkinson's disease (PD). There is an urgent need for developing biomarkers for diagnosing and monitoring the progression of Parkinson's disease. The combination of multiple cerebrospinal fluid biomarkers has emerged as an accurate diagnostic and prognostic model, while blood-based markers have

---

S. Mandal—Work was done while interning at IBM T. J. Watson Research Center.

C. Martín-Vide et al. (Eds.): AlCoB 2020, LNBI 12099, pp. 178–187, 2020.
https://doi.org/10.1007/978-3-030-42266-0_14

also been explored [10,22]. Recently, also gene expression and methylation signatures from blood samples have been examined for this purpose [30].

In the current work we examined the use of sequencing-based gene expression values from blood samples as features for predicting disease diagnosis. We found standard machine learning methods ineffective and instead looked into the possibilities of understanding the shape of the high-dimensional gene expression data by taking into account its topological features. More specifically, we examined persistent homology emerging from the gene expression data.

In life sciences, topological data analysis (TDA) has previously been applied in medical imaging [13,29], protein characterization [8,17], describing molecular architecture [24,26], and cancer genomics [3,21]. There have been several studies exploring TDA in genomics [7]. Gene expression from peripheral blood data has been used to build a model based on TDA network model and discrete Morse theory to look into routes of disease progression [27]. Persistent homology has also been employed for comparison of several weighted gene coexpression networks [18].

Here we describe a method that translates gene expression measurements for an individual sample into a weighted point cloud. We further hypothesize that this weighted point cloud has topological information relevant to classification tasks. Since the point cloud generated by directly mapping gene expressions can be large and in very high dimension, standard topological data analysis (TDA) algorithms suffer from a combinatorial explosion. We therefore employ subsampling and averaging methods using much fewer points. This subsampling method is also robust in terms of noise present in the original point cloud.

Ultimately, we use the gene expressions of subjects with and without Parkinson's disease to generate topological summaries per subject. These summaries essentially act as unique fingerprints that describe the topology of the gene expression in a sample. We use these fingerprints to enhance the feature vector that is used for disease phenotype prediction, and in turn achieve improved results compared to standard machine learning methods (support vector machines, random forests, neural networks). Our study also implies that gene expression measured from blood samples is a useful indicator of the presence or absence of Parkinson's disease.

## 2 Methods

### 2.1 Setting up the Problem

We work under the hypothesis that the set $X$ of all subjects' samples, each encoded as a collection of gene expression values, can provide us with enough topological information to discern between healthy subjects and subjects with Parkinson's disease. We denote by $X$ a matrix of size $n_{rows} \times n_{cols}$ where each row corresponds to a subject and each column corresponds to a gene. Each entry $X_{i,j}$ then corresponds to the $j$-th gene expression of the $i$-th subject.

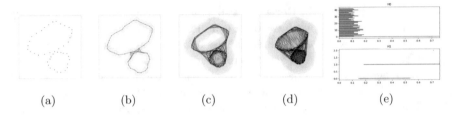

**Fig. 1.** Illustration of persistent homology. Here (e) represents the barcodes associated with increasing radius from (a) to (d).

The co-expression patterns between genes can reveal functional connections between them, e.g. the expression of a set of genes belonging to the same biological pathway may be co-ordinated (see [16] for applications of co-expression analysis). In particular, pathways perturbed in disease may contribute to gene expression differences between healthy and afflicted subjects.

Co-expression can be examined by computing pairwise correlations between gene expression measurements. Therefore, we construct a new matrix $\bar{X}$ from $X$, consisting of all pairwise distance correlations between *genes* (columns), rather than samples (rows), as defined in [28]. Distance correlation conveys different information about the relations between genes than the standard Pearson's correlation as it measures both linear and nonlinear association, whereas the former can only detect linear association.

Later, we show how to use the theory of persistent homology to determine the persistent topological landscapes present in the gene expression data of a sample, by first transforming it into a weighted point cloud. We do this transformation by utilizing the gene correlations across all available samples (matrix $\bar{X}$). The topological summaries of the weighted point clouds (persistence landscapes) are then used to construct a machine learning model to predict the phenotype (healthy or PD) for each sample.

## 2.2    Topological Background

In this section, we give a brief intuition as to how persistent homology is used to describe the shape of data. For a full account of basic constructions in this area, we refer the reader to [9]. We now give a brief intuitive description of a filtered Čech complex of a specific covering.

Consider the toy example of a set points sampled uniformly from a "figure eight" shape in $\mathbf{R}^2$ as shown in Fig. 1a. We can start growing disks around each point in the sample and consider the union of these disks. Initially when the radius $\rho$ of the disks are close to zero, we get a set of disconnected sets (Fig. 1a). As we continue, we notice that at some particular radius $\rho = \rho_1$, a set with the same topology as the figure eight is obtained (Fig. 1b). Upon increasing the radius further to $\rho = \rho_2$, the smaller of the two holes in the figure is completely filled while the other remains (Fig. 1c). Further increasing $\rho = \rho_3$, fills up the larger hole as well (Figure 1d), making the union of disks consist of a single connected component for all $\rho > \rho_3$.

The next step is to look at the combinatorial information contained in the evolving unions of disks. We achieve this by constructing a *filtered simplicial complex*, a mathematical object consisting of vertices, edges, triangles, tetrahedra, and their higher dimensional analogues, called simplices, with information on *when* these are added to the complex. First we add a vertex for each point in the point cloud at time 0, then whenever two disks intersect, we add an edge. Every time three disks intersect, we add a triangle, and similarly we add higher dimensional simplices for higher order intersections. Therefore, we get a sequence of simplices. In this sequence, the topological persistence computation is a set of *birth-death* pairs of homology cycle classes that indicate when a class is born and when it dies. In the previous example, it indicates when the loops are formed and when they are filled up. The pair (birth,death] of these homology classes can be indicated as a sets of points in $\mathbf{R}^2$ (called persistence diagrams; see [15]) or as sets of horizontal lines (called persistence barcodes; see [9]; illustrated in Fig. 1e).

Keeping track of the holes in this union of disks lets us know how long they persist, and we apply the same reasoning for more general point clouds with a notion of closeness or distance, even when they are not metrics in the mathematical sense. In particular we work with a set of points that form a *semimetric space*, a space that satisfies all but one of the metric spaces axioms: the triangle inequality does not hold in general.

In this article, we use the same construction as the one of a Weighted Vietoris-Rips complex [6, Section 5], pointing out that the properties obtained in the cited article state as hypothesis that the input space is metric. It is obvious, however, that applying the construction to a semimetric space still yields a filtered simplicial complex.

The rank of homology groups of a space, in the usual non-persistent setting, are called Betti numbers of said space, and are denoted by $\beta_k - number$. Intuitively, these numbers account for different topological features: $\beta_0$ indicates the number of connected components, $\beta_1$ indicates the number of cycles or loops, whereas $\beta_2$ counts the number of voids. In our example, Fig. 1e represents barcodes associated with Figs. 1a–d. In Fig. 1e ($H_0$), we have many $\beta_0$ at the beginning, as observed in Fig. 1a which die as $\rho$ increases to connect the different components. In Fig. 1e ($H_0$) we have a red bar, which corresponds to the final connected component that lives forever. In Fig. 1e ($H_1$), we notice two long bars, representing the two loops that are formed and then eventually filled up. The fact that one bar is longer than the other suggests the size of one hole greater than the other. Also, notice that these bars in $H_1$ appear up after the cycles in $H_0$ have died. In our data analysis using topological features, we wish to use these intervals in persistent homology to better understand the shape of the data.

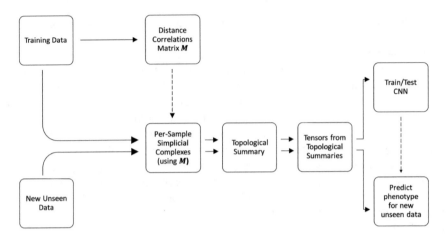

**Fig. 2.** TDA pipeline flowchart.

## 2.3 TDA Workflow

Our overall workflow is described in Fig. 2 and the relevant steps are discussed in detail below.

Consider $Z = X_{train}$, a matrix of size $m_{rows} \times n_{columns}$ corresponding to a subset of samples in $X$ that will be used to train a classifier later on. Recall that $Z_{i,j}$ corresponds to the $j$-th gene expression of the $i$-th subject in the gene expression matrix $Z$, and that the matrix $\overline{Z}$ consists of all pairwise sample distance correlations between *columns* (genes) of $Z$. From [28, Theorem 3], we know that the distance correlation between two vectors (samples of a random variable) is zero if and only if are they are independent, that it is non-negative, and that it is bounded above by 1. It is also symmetric. In our particular data, we additionally have that no two different columns have distance correlation of 1.

Next we define the matrix $M = \mathbf{1} - \overline{Z}$, where $\mathbf{1}$ denotes the $m_{rows} \times n_{cols}$ constant matrix with value 1. By the preceding paragraph, $M$ defines now a semimetric on the set of genes in our dataset.

We now, for each subject $s$ in $X$, construct a filtered simplicial complex $K_s$, as follows:

1. Its vertices are the set of genes. All added at time 0.
2. For each edge $\sigma = (g_i, g_j)$, we compute

$$t_\sigma = \frac{\sqrt{M_{i,j}^2 + \left(\frac{X_i^2 - X_j^2}{M_{i,j}}\right)^2 + 2X_i^2 + 2X_j^2}}{2},$$

and add $\sigma$ to $K_s$ at time $t_\sigma$.
3. For each simplex $\sigma$ with $|\sigma| > 2$ we add it at the maximum time of addition of its edges.

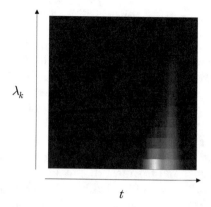

**Fig. 3.** Discrete sampling of a persistence landscape of a subject, brighter colors indicate higher values in the landscape.

This is an analytical solution for adding simplices to the Weighted Vietoris-Rips construction, studied in [6], for the semi metric space $M$ (of genes) with weights on $X$, the original data matrix. We are essentially assigning to each gene its expression associated to the subject $s$. In the actual implementation, we multiplied the weights by a scaling term, since all distances are bound by 1 above but the weights themselves can be higher.

To mitigate the computational cost of our setup we used a subsampling approach, as studied in [11], so that instead of working with the entire set of genes at all times, for each subject we repeatedly subsampled smaller sets of $n_{subsample}$ genes, obtaining several filtered simplicial complexes.

For each of the simplicial complexes we obtained persistence landscapes [5] for homology dimensions 0 and 1. Such landscapes are, for each homology degree, sequences $\{\lambda_k\}$ of decreasing piecewise linear (PL) functions $\lambda_k : \mathbb{R} \to \mathbb{R}$. We elected persistence landscapes as opposed to barcodes or diagrams because of their amenability to compute statistical estimators, such as averages. After computation of all landscapes, for each subject we then obtained its average landscape.

We then quantized the resulting landscapes by sampling $r_x$ values evenly in the interval $[0, t_{max}]$, where $t_{max}$ is a value estimated from the data that corresponds to the last time of the filtration where there were changes in the persistent homology of the complex being processed. This results, for each persistence landscape, in a 2D array of size $r_x \times n_\lambda$, where $n_\lambda$ is the number of non-zero PL functions in the landscape. See Fig. 3 for an example of one such landscape.

Note that other alternatives to performing vectorization of topological summaries, such as persistence images [2], exist and could be used in place of persistence landscapes, but to compare their performance is beyond the scope of this article.

## 2.4    Machine Learning Framework

For each subject, we obtained a feature vector of 19,581 gene expression measurements (see Sect. 2.6) and a known class label 0 (161 control subjects) or 1 (264 affected subjects) according to the Parkinson's disease phenotype. We split the data 80-20 into training and test sets, over 50 iterations, except for the computationally more intensive TDA-CNN where we considered 4 iterations after observing the results between iterations were nearly identical.

First we generated a basis of comparison for our TDA approach (TDA-CNN) with standard machine learning algorithms. We used several widely used binary classifiers to train a model and then test its predictions on unseen data: support vector machines with radial basis function kernel (SVM-RBF) and linear kernel (SVM-Linear), as well as random forest (RF) and a simple neural network (MLP-NN, consisting of 3 hidden layers with 20 neurons each, using relu activations). These methods were applied using the scikit-learn python library [23].

In the TDA-CNN approach, for a given resolution $r_y$, we fed each subject's vectorized persistence landscape as a tensor of shape $(r_x, r_y, 2)$, one channel per homology degree, into a Convolutional Neural Network (CNN) [20] implemented using the Keras library [12] with the Tensorflow [1] backend. We employed a nearest neighbors filter to scale in the $y$-axis when $n_\lambda \neq r_y$. The architecture we used consisted of two separate paths, one per channel, each consisting of 3 convolutional layers, each with 64 $3 \times 3$ filters, with Max-Pooling layers [25] of size $2 \times 2$ after each convolutional layer. We then fed the outputs of these two paths into a dense layer with 32 neurons. Finally we used a 2-neuron layer with softmax activations [4] as output. All other activations were exponential linear units [14].

## 2.5    Topological Data Analysis Implementation

We used our own implementation, **maTilDA: Multi-Purpose Toolkit for TDA**, for the construction of the Weighted Vietoris-Rips complex, its persistent homology, barcode computations, persistence landscapes and discrete sampling.

## 2.6    Gene Expression Data Processing

We downloaded gene expression data derived from RNA sequencing, acquired from blood samples of Parkinson's disease (PD) and control subjects, from the Parkinson's Progression Markers Initiative (https://www.ppmi-info.org/, Phase 1 data). The downloaded sequencing read counts per gene (1,889 samples and 57,820 genes) were examined and outliers removed: 1,141 abundantly expressed RN7S cytoplasmic genes and 742 samples whose read count distributions were different from other samples (having reads assigned to more than 35k genes) were removed.

**Table 1.** $F_1$ scores (micro and macro), true positive rate (TPR), and true negative rate (TNR) for predicting Parkinson's disease status from gene expression data. Underlined numbers indicate best values per row.

|  | TDA-CNN | SVM-Linear | SVM-RBF | Random Forest | MLP-NN |
|---|---|---|---|---|---|
| $F_1$-micro | 0.877 | 0.629 | 0.641 | 0.641 | 0.556 |
| $F_1$-macro | 0.871 | 0.580 | 0.477 | 0.549 | 0.400 |
| TPR | 0.870 | 0.773 | 0.960 | 0.874 | 0.672 |
| TNR | 0.890 | 0.391 | 0.112 | 0.256 | 0.362 |

RoDEO [19] was applied (with parameters $P = 20$ bins, $I = 10$ iterations, and $R = 10^8$ reads) to scale the samples to a normalized range of counts $[1, 20]$. For the study in this paper we focused on $20,345$ protein-coding genes per sample. We further removed uninformative genes with a constant RoDEO projected value for each sample, and duplicate genes with identical value distributions, obtaining a final set of $19,581$ genes.

For the purpose of this study, subjects with Parkinson's disease (cohorts PD, GENPD, REGPD) were denoted by phenotype label 1, and subjects without the disease (HC, GENUN, REGUN) by phenotype label 0. We thus had a total of 264 PD samples and 161 unaffected samples (if a subject had several time points sequenced, we took the first one).

## 3   Results and Discussion

Table 1 shows the $F_1$ score, true positive rate (TPR), and true negative rate (TNR) for our TDA approach as well as for baseline machine learning methods, on the task of predicting Parkinson's disease diagnosis. We included both micro-$F_1$ and macro-$F_1$ scores, the latter gives equal weight to each class to take into account class imbalance [31]. Our TDA with the convolutional neural network (TDA-CNN) approach achieved remarkable improvement of both micro- and macro-$F_1$ scores compared to the other methods. The TDA-CNN approach achieve scores above 0.87, while SVM (with two different kernel options) and random forest yield similar values to each other, up to 0.64 for micro-$F_1$ and 0.58 for macro-$F_1$. Note that convolutional neural networks operate on two-dimensional image data, thus we chose a multilayer perceptron neural network (MLP-NN) as a comparison on the RoDEO-processed gene expression vector data. For this data, the MLP-NN approach seems particularly ill suited, with scores 0.56 and 0.40 for micro- and macro-$F_1$, respectively.

In terms of true positive and true negative rates, the standard methods have very low TNR, indicating abundant false positives. The TDA-CNN approach, on the other hand achieves a balance between sensitivity and specificity with TPR and TNR having high values (0.87 and above).

The findings indicate blood-based gene expression does contain signal that is relevant for separating subjects with and without Parkinson's disease. Further work includes understanding these subtle signals in order to transform the findings into diagnostic and prognostic models. The introduced framework of topological data analysis with convolutional neural network prediction is a general approach that could be applied to gene expression data relating to other phenotypes.

**Acknowledgments.** Saugata Basu was partially supported by NSF Grant DMS-1620271.

# References

1. Abadi, M., et al.: TensorFlow: large-scale machine learning on heterogeneous systems (2015). https://www.tensorflow.org/. Software available from tensorflow.org
2. Adams, H., et al.: Persistence images: a stable vector representation of persistent homology. J. Mach. Learn. Res. **18**(8), 1–35 (2017). http://jmlr.org/papers/v18/16-337.html
3. Arsuaga, J., Borrman, T., Cavalcante, R., Gonzalez, G., Park, C.: Identification of copy number aberrations in breast cancer subtypes using persistence topology. Microarrays **4**(3), 339–369 (2015)
4. Bridle, J.S.: Probabilistic interpretation of feedforward classification network outputs, with relationships to statistical pattern recognition. In: Soulié, F.F., Hérault, J. (eds.) Neurocomputing. NATO ASI Series (Series F: Computer and Systems Sciences), vol. 68, pp. 227–236. Springer, Heidelberg (1990). https://doi.org/10.1007/978-3-642-76153-9_28
5. Bubenik, P.: Statistical topological data analysis using persistence landscapes. J. Mach. Learn. Res. **16**(1), 77–102 (2015). http://dl.acm.org/citation.cfm?id=2789272.2789275
6. Buchet, M., Chazal, F., Oudot, S.Y., Sheehy, D.R.: Efficient and robust persistent homology for measures. Comput. Geom. Theory Appl. **58**(C), 70–96 (2016). https://doi.org/10.1016/j.comgeo.2016.07.001
7. Camara, P.: Topological methods for genomics: present and future directions. Curr. Opin. Syst. Biol., 95–101 (2017). https://doi.org/10.1016/j.coisb.2016.12.007
8. Cang, Z., Mu, L., Wu, K., Opron, K., Xia, K., Wei, G.W.: A topological approach for protein classification. Comput. Math. Biophys. **3**(1), 140–162 (2015). https://doi.org/10.1515/mlbmb-2015-0009
9. Carlsson, G., Zomorodian, A., Collins, A., Guibas, L.: Persistence barcodes for shapes. In: Proceedings of the 2004 Eurographics/ACM SIGGRAPH Symposium on Geometry Processing. SGP 2004, pp. 124–135. ACM, New York (2004). https://doi.org/10.1145/1057432.1057449
10. Chahine, L.M., Stern, M.B., Chen-Plotkin, A.: Blood-based biomarkers for Parkinson's disease. Parkinsonism Relat. Disord. **20**(S1), S99–S103 (2014)
11. Chazal, F., Fasy, B., Lecci, F., Michel, B., Rinaldo, A., Wasserman, L.: Subsampling methods for persistent homology. In: Bach, F., Blei, D. (eds.) Proceedings of the 32nd International Conference on Machine Learning. Proceedings of Machine Learning Research, vol. 37, pp. 2143–2151. PMLR, Lille, France, 07–09 July 2015. http://proceedings.mlr.press/v37/chazal15.html
12. Chollet, F., et al.: Keras (2015). https://keras.io
13. Chung, M.K., Bubenik, P., Kim, P.T.: Persistence diagrams of cortical surface data. In: Prince, J.L., Pham, D.L., Myers, K.J. (eds.) IPMI 2009. LNCS, vol. 5636, pp. 386–397. Springer, Heidelberg (2009). https://doi.org/10.1007/978-3-642-02498-6_32
14. Clevert, D.A., Unterthiner, T., Hochreiter, S.: Fast and accurate deep network learning by exponential linear units (elus). arXiv:1511.07289 (2015)

15. Cohen-Steiner, D., Edelsbrunner, H., Harer, J.: Stability of persistence diagrams. In: Proceedings of the Twenty-first Annual Symposium on Computational Geometry. SCG 2005, pp. 263–271. ACM, New York (2005). https://doi.org/10.1145/1064092.1064133

16. van Dam, S., Võsa, U., van der Graaf, A., Franke, L., de Magalhães, J.P.: Gene co-expression analysis for functional classification and gene-disease predictions. Brief. Bioinform. **19**(4), 575–592 (2017). https://doi.org/10.1093/bib/bbw139

17. Dey, T., Mandal, S.: Protein classification with improved topological data analysis. In: 18th International Workshop on Algorithms in Bioinformatics (WABI 2018). Leibniz International Proceedings in Bioinformatics (2018)

18. Duman, A.N., Pirim, H.: Gene coexpression network comparison via persistent homology. Int. J. Genomics **2018**, Article ID 7329576, 1–11 (2018). https://doi.org/10.1155/2018/7329576

19. Haiminen, N., et al.: Comparative exomics of Phalaris cultivars under salt stress. BMC Genomics (Suppl 6), S18 (2014). https://doi.org/10.1186/1471-2164-15-S6-S18

20. Le Cun, Y., et al.: Handwritten digit recognition with a back-propagation network. In: Proceedings of the 2nd International Conference on Neural Information Processing Systems. NIPS 1989, pp. 396–404. MIT Press, Cambridge (1989). http://dl.acm.org/citation.cfm?id=2969830.2969879

21. Nicolau, M., Levine, A.J., Carlsson, G.: Topology based data analysis identifies a subgroup of breast cancers with a unique mutational profile and excellent survival. Proc. Natl. Acad. Sci. **108**(17), 7265–7270 (2011). https://doi.org/10.1073/pnas.1102826108

22. Parnetti, L., et al.: CSF and blood biomarkers for Parkinson's disease. Lancet Neurol. **18**(6), 573–586 (2019)

23. Pedregosa, F., et al.: Scikit-learn: machine learning in Python. J. Mach. Learn. Res. **12**, 2825–2830 (2011)

24. Pike, J.A., et al.: Topological data analysis quantifies biological nano-structure from single molecule localization microscopy. bioRxiv (2018). https://doi.org/10.1101/400275

25. Ranzato, M., Huang, F.J., Boureau, Y., LeCun, Y.: Unsupervised learning of invariant feature hierarchies with applications to object recognition. In: 2007 IEEE Conference on Computer Vision and Pattern Recognition, pp. 1–8, June 2007. https://doi.org/10.1109/CVPR.2007.383157

26. Sauerwald, N., Shen, Y., Kingsford, C.: Topological data analysis reveals principles of chromosome structure throughout cellular differentiation. bioRxiv (2019). https://doi.org/10.1101/540716

27. Schofield, J.P.R., et al.: A topological data analysis network model of asthma based on blood gene expression profiles. bioRxiv (2019). https://doi.org/10.1101/516328

28. Székely, G.J., Rizzo, M.L., Bakirov, N.K.: Measuring and testing dependence by correlation of distances. Ann. Stat. **35**(6), 2769–2794 (2007). https://doi.org/10.1214/009053607000000505

29. Turner, K., Mukherjee, S., Boyer, D.M.: Persistent homology transform for modeling shapes and surfaces. Inf. Infer. **3**(4), 310–344 (2014)

30. Wang, C., Chen, L., Yang, Y., Zhang, M., Wong, G.: Identification of potential blood biomarkers for Parkinson's disease by gene expression and DNA methylation data integration analysis. Clin. Epigenetics **11**, 24 (2019)

31. Yang, Y., Liu, X.: A re-examination of text categorization methods. In: Proceedings of the 22nd Annual International ACM SIGIR Conference on Research and Development in Information Retrieval, Berkeley, 15–19 August 1999, pp. 42–49. ACM (1999)

# BOAssembler: A Bayesian Optimization Framework to Improve RNA-Seq Assembly Performance

Shunfu Mao[(✉)] , Yihan Jiang , Edwin Basil Mathew ,
and Sreeram Kannan

University of Washington, Seattle, WA 98195, USA
{shunfu,yij021,edwin100,ksreeram}@uw.edu

**Abstract.** High throughput sequencing of RNA (RNA-Seq) can provide us with millions of short fragments of RNA transcripts from a sample. How to better recover the original RNA transcripts from those fragments (RNA-Seq assembly) is still a difficult task. For example, RNA-Seq assembly tools typically require hyper-parameter tuning to achieve good performance for particular datasets. This kind of tuning is usually unintuitive and time-consuming. Consequently, users often resort to default parameters, which do not guarantee consistent good performance for various datasets.

**Results:** Here we propose BOAssembler, a framework that enables end-to-end automatic tuning of RNA-Seq assemblers, based on Bayesian Optimization principles. Experiments show this data-driven approach is effective to improve the overall assembly performance. The approach would be helpful for downstream (e.g. gene, protein, cell) analysis, and more broadly, for future bioinformatics benchmark studies.

**Availability:** https://github.com/shunfumao/boassembler.

**Keywords:** RNA-Seq · Assembly · Bayesian Optimization

## 1 Introduction

Sequence assembly is a process to recover the original genomic sequences from their sampled reads. Based on sequence type (DNA/RNA) and the availability of reference genome, there are different assembly problems. In this study, we focus on reference-based RNA-Seq assembly, which is a critical step to understand gene, protein and cell functions.

Existing popular reference-based RNA-Seq assemblers include Cufflinks [3] and Stringtie [2]. They usually align reads onto reference genome first, and utilize the read alignments to build a graph where each node represents a genome region (exon) and each edge represents the connection between two nodes by some

---

S. Mao and Y. Jiang—Equal contributors.

C. Martín-Vide et al. (Eds.): AlCoB 2020, LNBI 12099, pp. 188–197, 2020.
https://doi.org/10.1007/978-3-030-42266-0_15

reads. They then traverse the graph to find paths as the reconstructed RNA transcripts.

These assembly problems are essentially NP-hard [7] and existing tools resort to heuristic methods. For example, from the graph, Stringtie will extract the heaviest paths iteratively. Due to the heuristic approaches, these methods usually require parameter tuning to achieve good performance for particular datasets. Since most users may not understand the meaning of the parameters well and tuning itself is tedious and time-consuming, they usually end up with default settings. An automatic tuning framework, therefore, is necessary.

In machine learning (ML), Bayesian Optimization (BO) is gaining a surge of interest as its usefulness in tuning hyper-parameters for modern deep learning systems [10, 11]. BO is favorable for optimizing objective functions that are expensive to evaluate and are over continuous domains of less than 20 dimensions [12]. BO has been widely used in most deep learning systems such as Natural Language Processing (NLP) [13], Reinforcement Learning (RL) [14], and Channel Coding [15]. Depending on algorithms and programming languages, several popular BO packages have been developed, such as GPyOpt [16].

There are limited work to introduce BO into computational biology fields. Recently [17] applies BO to improve eQTL analysis. To the best of our knowledge, no work has introduced BO to assembly tasks yet, which are fundamentally graph problems with their own unique challenges. To fill this gap, we have developed BOAssembler, which is a framework able to incorporate existing assemblers (such as Stringtie) and BO methods (such as GPyOpt) to assist assembler developers and biologists to spend minimal efforts to obtain better assembly hyper-parameters automatically fine-tuned for particular datasets.

Our contributions include: (a) We firstly explore the BO methods in (reference-based RNA-Seq) assembly tasks. (b) Our designed experiments show that BO is overall effective to improve assembly. (c) An open source end-to-end framework (BOAssembler) is provided for the assembly community to use.

## 2    Methods

### 2.1    Assembly

There are two kinds of RNA-Seq assembly problems: de novo assembly and reference-based assembly. For de novo assembly, we only have RNA-Seq reads, which is common in non-model organisms. For reference-based assembly, there is additional knowledge on the genome of the organism. De novo assembly is apparently more challenging and typical tools (such as Trinity [4] and recently Shannon [5]) require much more computational resources and more complicated evaluations. As the first step to bridge assembly and BO, we focus on reference-based RNA-Seq assembler. In particular, we focus on the widely used Stringtie, as recommended in [6].

A typical reference-based RNA-Seq assembly includes aligning sampled RNA-Seq reads onto a reference genome using external tools such as STAR [1] etc. For Stringtie, a (splice) graph will be prepared where each node represents a unique exonic region supported by aligned reads and edges indicate how nodes are bridged by reads. Graph traversal algorithms will be applied to find paths as transcripts to best explain the constraints from graph nodes and edges.

Since assembly problems are NP hard [7], existing algorithms take a lot of heuristics (predefined threshold values). Therefore, assembler performance heavily depends on its parameters. For example in Stringtie, the parameter '-f' sets a fractional threshold so that the predicted transcripts having a lower relative abundance level than this will be discarded; a reduced '-f' threshold therefore encourages transcripts to be retained to improve sensitivity.

Developers of assemblers typically tune parameters by intuition on a few datasets, and offer selected parameters for assembler users to use. As the assembly performance for various datasets are usually parameter dependent, a more systematic method of tuning parameter is needed. Naive approaches would include grid search or random search. However, because typical assembly parameters are continuous, and of around 10 to 20 dimensions, which could make grid search on all possible combinations prohibitive. Random search [9], on the other hand, are expensive to guarantee good coverage.

## 2.2   Bayesian Optimization

BO aims at maximizing a real-value black-box function $f(\theta)$ with respect to $\theta$ [8] in a gradient-free approach. BO consists of a statistical surrogate objective function to model the input-output relationship between $\theta$ and $f(\theta)$, and an acquisition function to decide what to sample next. Firstly BO evaluates randomly chosen $K$ datapoints of $\theta$, and fits the prior statistical objective model. Then BO iteratively updates the posterior model with newly acquired $f(\theta_k)$, and selects $\theta_{k+1}$ to evaluate according to posterior. BO is a systematic approach to explore the parameter space according to a Bayesian model with limited allowed evaluations (i.e. iterations).

Our BOAssembler primarily uses Gaussian Process (GP) with Matern Kernel [11] as a natural model for statistical objective function, and Expected Improvement (EI) as a commonly used acquisition function. Specifically, assume we want to sample a new datapoint $\theta$, and our current best parameter is $\theta^*$. Then the improvement is defined as $[f(\theta) - f(\theta^*)]^+$. Note that the improvement is positive only when $f(\theta)$ is larger than $f(\theta^*)$. Then the expected improvement can be taken under posterior distributions of $f$ given $\theta_{1:k}$:

$$EI_k(\theta) = \mathbb{E}([f(\theta) - f(\theta^*)]^+ | \theta_{1:k}, f(\theta_{1:k})) \tag{1}$$

As the expected improvement can be computed in closed-form, we can select the point with largest expected improvement to sample: $\theta_{k+1} = \arg \max EI_k(\theta)$.

The procedure of iterative update, based on GP and EI, is described in Algorithm 1, which includes $K$ datapoints for BO initialization, and $T$ iterations to acquire.

**Input:**$D$, $K$, $T$
**Output:**Best parameter $\theta^*$
Fit the GP with $K$ initial samples $\theta_k, k \in \{1, ..., K\}$;
i=0;
**while** $i < T$ **do**
    Update the GP posterior probability distribution on $f$ using all available data;
    Use EI to compute the $\theta_{i+1}$ with updated posterior distribution;
    Obtain $f(\theta_{i+1})$;
    $i = i + 1$;
**end**
Return $\theta^*$ with best performance;

**Algorithm 1.** Baysian Optimization Algorithm

## 2.3   Combine BO and Assembly

**The Motivation to Combine.** We can formulate the problem of assembly parameter tuning as follows. The reference-based assembler together with its performance evaluation can be represented as an abstract function $f(D, \theta)$, where $D$ includes both the read alignments used for assembly and the reference transcriptome (a set of ground truth RNA transcripts) used for evaluation, and $\theta$ refers to the parameters of $f$. After read alignments are assembled with given parameter $\theta$, the assembly output (a set of RNA transcripts) will be compared with the reference transcriptome, and the quality of assembly is measured by scalar metrics such as precision $p$ and sensitivity $s$. $f(D, \theta)$ outputs an evaluation score based on $p$ and $s$. Our goal is to find a global optimal $\theta$ which maximizes $f(D, \theta)$ under limited number of iterations, as running assembler per iteration is time consuming.

Bayesian optimization works well for black-box gradient-free global optimization with moderate dimensionality. It is thus favorable to apply BO to optimize assembler parameters due to the following reasons: (a) Empirically BO works well for parameters with moderate dimensionality (less than 20). This is consistent with assemblers which typically have this number of parameters. (b) $f(.)$ is continuous, and the parameter $\theta$ are correlated, and has well-defined feasible set. The assembler parameters usually have continuous values within certain ranges. (c) The function $f(.)$ is expensive to evaluate, thus to evaluate all possible combinations of parameters is prohibitive. Indeed, assembly tasks are time consuming. (d) $f(.)$ is a 'black-box', while gradient-based optimization methods cannot be applied. Assemblers typically do not have a gradient due to the usage of thresholds, which makes black-box method favorable.

**The Architecture.** Figure 1 illustrates the overall architecture of BOAssembler. There are two parts: the assembly part (e.g. $f(D, \theta)$) and BO part.

The assembly part wraps up the RNA-Seq reference-based assembler (here Stringtie), which takes fixed read alignments as well as adjustable assembler

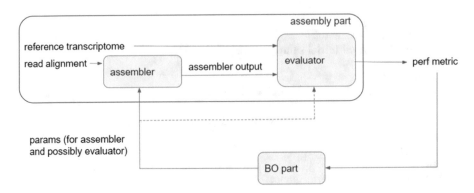

**Fig. 1.** BOAssembler architecture.

parameters as input, and outputs assembled RNA transcripts (in gtf format). In addition, the assembly part includes an evaluator block to access the assembly output. Basically, it calls the gffcompare[1] tool, which takes as input the assembly output and reference transcriptome, and outputs sensitivity and precision statistics. The sensitivity is the percentage of reference RNA transcripts that have been correctly recovered, and the precision means the percentage of assembled transcripts that correctly match the reference transcriptome. We further combine the sensitivity and precision (such as F1 score) as $f(D,\theta)$, to be used by the BO part. The evaluator may also take adjustable parameters as discussed in Sect. 2.3.

The BO part has its theory described in Sect. 2.2. The BO part mainly relies on GPyOpt, which implements the core BO methods (e.g. the GP + EI approach). The BO part treats the assembly part as a black box, where the input to the box is the parameters for assembler and evaluator, and the output of the box is the combined performance metric for the assembler (such as F1 score). The BO part will iteratively optimize the parameters for the black box function (e.g. the assembly part) based on the feedback of performance metric.

**Metrics to Optimize.** The assembly part outputs $f(D,\theta)$, which is a metric score and serves as an input to BO part. In particular, it is defined as a weighted F1 score $(S_w = \frac{\lambda p \times (1-\lambda)s}{\lambda p + (1-\lambda)s})$ on top of the evaluator's output in terms of sensitivity $s$ and precision $p$ of the assembly. $\lambda \in (0, 0.5)$ is also BO tunable.

There are several candidate metrics including the mean value $(S_m = \frac{s+p}{2})$ and the F1 score $(S_{F1} = \frac{2sp}{s+p})$. We found the BO part tends to overfit either $s$ or $p$ towards 1 when using $S_m$. Though $S_{F1}$ is able to balance $s$ and $p$, we find $S_w$ is better to improve the final performance of sensitivity and precision. In our experiments, $s$ tends to have a lower value range than $p$ (due to many reference RNA transcripts do not have enough coverage), we hope to reward more for sensitivity improvement but still have gain on precision. Therefore, we

---

[1] https://ccb.jhu.edu/software/stringtie/gffcompare.shtml.

come up with the weighted F1 score $S_w$, which uses BO to figure out how much percentage we want to reward especially for the improvement of sensitivity.

**Hyper-parameters.** Our final goal is to find good assembler parameters. To achieve this, we hope to find hyper-parameters ($\theta$) that achieves high metric score $S_w$. $\theta$ include both assembler parameters and evaluator parameter ($\lambda$). Each parameter has its name, type, default value and range. For example, Stringtie has a parameter '-f' with type float, default value 0.1 and range $(0.0, 1.0)^2$.

**Usage and Extension of BOAssembler.** To tune assembler parameters for a particular dataset in BOAssembler, the user only needs to provide a small sample of read alignments of the target dataset. The sampling can be done by our provided scripts. After some iterations, BOAssembler will report suggested parameters and its tuning history.

BOAssembler currently uses Stringtie as its default assembler. It supports Cufflinks as well. Extension to use other reference-based RNA-Seq assemblers is also straightforward. The user only needs to follow the Stringtie example, to add a line of Python code in a specified Python file, and to add a config file which contains the parameters to be tuned.

## 3   Result and Discussion

### 3.1   Datasets

Our goal is to use BOAssembler to tune assembler's hyper-parameters on a smaller dataset, and apply recommended hyper-parameters on a large assembly task. Since the smaller dataset has representative data of large assembly task, we expect tuned hyper-parameters can overall improve the large assembly task in terms of sensitivity and precision.

We build our results based on simulated datasets, since real datasets lack ground truth and it is hard to judge if an assembled RNA transcript is a false positive, or a new RNA transcript that has yet to be discovered. Consequently, the evaluator metric $S_w$ is not feasible for real datasets. The simulated datasets are generated based on real ones in the following steps.

Firstly we prepare three real datasets, including: 132.05M Illumina single end reads (50-bp) sampled from human embryonic stem cells (HESC) (GSE51861, used in [18]), 115.36M Illumina pair end reads (101-bp) sampled from Lymphoblastoid cells (LC) (SRP036136, used in [19]), and 183.53M Illumina pair end reads (100-bp) sampled from HEK293T (Kidney) cells (SRX541227), previously produced and studied in StringTie [2].

Secondly, we use RSEM [20] to generate simulated reads from real datasets. To begin with, we choose LC reference transcripts (containing 207266 RNA

---

$^2$ See http://ccb.jhu.edu/software/stringtie/index.shtml?t=manual for an example of Stringtie's parameters.

transcripts) as the ground truth reference transcriptome annotations. We then do quantification of real datasets using RSEM and get learned statistics from real datasets. Based on learned statistics, we use RSEM to sample simulated reads from ground truth reference transcriptome. The simulated HESC has 150M 50-bp single-end reads, the simulated LC dataset has 150M 101-bp pair end reads and the simulated Kidney dataset has 150M 100-bp pair end reads.

Lastly, we use STAR [1] (2-pass strategy) to align three simulated datasets onto the human reference genome (hg19)[3]. From each alignment (in bam format), we subsample to get smaller alignment files of chromosome15 as fixed datasets for BOAssembler. The small datasets are about 1.5%, 3.1%, and 2.1% of large datasets for HESC, LC and Kidney respectively. We've proposed another more complicated sampling method (available at Github site) across chromosomes, which offers similar performance.

## 3.2    Experiment Procedure

For each dataset, we run BOAssembler on the smaller datasets. The evaluation for metric also uses a subset (e.g. chromosome 15) of reference transcriptome. Each iteration takes around 1 min, and we typically see convergence of metric score around 40 to 50 iterations. Compared to grid search for possible combinations of 10 to 20 parameters, BOAssembler is much more efficient.

After automatic tuning, BOAssembler will recommend parameters with high metric scores. We then apply these parameters on large datasets, which typically take several hours to finish the assembly tasks using 25 cores of a linux server.

## 3.3    Experiment Results

Table 1 compares the performance of default parameters (Default) and BOAssembler-tuned parameters (Tuned) for each simulated dataset, in terms of sensitivity and precision. We also list their standard F1 score here since it's related to the metric BOAssembler tries to optimize. But we'll focus on sensitivity and precision which are of practical interest.

As Table 1 shows, BOAssembler has improved sensitivity, and precision for all small datasets. In particular, HESC small is improved by 16.9% in sensitivity and 27.4% in precision, LC small is improved by 1.2% in sensitivity and 13.1% in precision, Kidney small is improved by 3.2% in sensitivity and 5.5% in precision. Notice that the real Kidney dataset has been used in Stringtie's original work, so the default parameters of Stringtie should have been adjusted for this dataset statistics. Still BOAssembler improves the its performance further.

The trend of performance improvement is mostly reflected in assembly tasks on large datasets, which is mostly interesting to us. In particular, HESC large is improved by 6.5% in sensitivity and 32.2% in precision, Kidney large is improved by 0.3% in sensitivity and 0.6% in precision. LC large has a small loss around

---

[3] http://hgdownload.cse.ucsc.edu/goldenpath/hg19/bigZips/.

**Table 1.** Performance on different simulated datasets

| Dataset | Sensitivity (%) | | Precision (%) | | F1 | |
|---|---|---|---|---|---|---|
| | Default | Tuned | Default | Tuned | Default | Tuned |
| HESC (small) | 22.1 | 39 | 31.9 | 59.3 | 26.11 | 47.05 |
| HESC (large) | 14.3 | 20.8 | 54.2 | 86.4 | 22.63 | 33.53 |
| LC (small) | 25.8 | 27 | 40 | 53.1 | 31.37 | 35.8 |
| LC (large) | 15.7 | 14.5 | 64.3 | 74.3 | 25.24 | 24.26 |
| Kidney (small) | 20.1 | 23.3 | 27.9 | 33.4 | 23.37 | 27.45 |
| Kidney (large) | 14.8 | 15.1 | 54.1 | 54.7 | 23.24 | 23.67 |

1.2% in sensitivity, but it gains 10%, which is significant, in precision. The experiments show that by tuning hyper-parameters through BOAssembler on small datasets, we are able to improve large assembly tasks overall (though there could be fluctuations) to a smaller extent.

The diminished performance gain of tuned parameters on large datasets, compared to the gain on small ones, may be because of an averaging effects across more variant alignment statistics in large datasets. To better catch up large dataset statistics, we have also prepared small datasets selected from certain regions, the performance improvement trend is similar.

By comparing the BOAssembler suggested parameters with assembler's default ones, we could also gain more insights into the datasets. For example, in HESC small datasets, the parameter 'f' is suggested to decrease from 0.1 to 0, this will allow more transcripts of low expression levels to also be considered as assembly output (hereby improve sensitivity). Meanwhile, the parameter 'm' is suggested to increase from 200 to 500 to allow only longer (e.g. at least 500) assembled transcripts to be considered (hereby improve precision).

### 3.4   Discussion

We expect our study and developed BOAssembler will contribute to the assembly community as follows: (a) For bioinformaticians who develop assembly algorithms, the framework or ideas behind it could provide them with more convenient ways to set default parameters for their assemblers. (b) For biologists who use reference-based RNA-Seq assemblers, BOAssembler can help them improve assembly performance, so they can gain better insights into the datasets, and the improved assembled RNA transcripts will be helpful for downstream gene, protein and cell related analysis. (c) For benchmark work of assemblers, typically several datasets are prepared and different assemblers are compared by using their default parameters. BOAssembler or its ideas will help the benchmark work in a fairer basis, since default parameters can not guarantee consistent good performance across various datasets.

Whereas this is, to our best knowledge, the first efforts to bring assembly and BO together, there are several interesting future directions. (a) As from experiments, we have observed that the gain of tuned parameters gets diminished for larger datasets, which implies BO tuned parameter overfits to small training dataset. Since evaluating assembler is expensive, more efficient data subsampling and cross-validation methods to avoid overfitting will be helpful. (b) Another interesting exploration is how to define a metric score that is better than the current weighted F1 score for Bayesian Optimization, to better balance sensitivity and precision. (c) There're many problems in assembly areas (including variant calling) that heavily relay on hyper-parameter tuning for better performance. Introduce similar frameworks to these problems shall have wide applications.

**Funding.** This project is funded by NIH R01 Award 1R01HG008164 by NHGRI and NSF CCF Award 1703403.

# References

1. Dobin, A., et al.: STAR: ultrafast universal RNA-seq aligner. Bioinformatics **29**(1), 15–21 (2013). https://academic.oup.com/bioinformatics/article/29/1/15/272537
2. Pertea, M., Pertea, G.M., Antonescu, C.M., Chang, T.-C., Mendell, J.T., Salzberg, S.L.: StringTie enables improved reconstruction of a transcriptome from RNA-seq reads. Nat. Biotechnol. **33**, 290–295 (2015)
3. Trapnell, C., et al.: Transcript assembly and quantification by RNA-seq reveals unannotated transcripts and isoform switching during cell differentiation. Nat. Biotechnol. **28**(5), 511–515 (2010)
4. Grabherr, M.G., et al.: Full-length transcriptome assembly from RNA-seq data without a reference genome. Nat. Biotechnol. **29**(7), 644–652 (2011)
5. Kannan, S., Hui, J., Mazooji, K., Pachter, L., Tse, D.: Shannon: an information-optimal de Novo RNA-seq assembler. bioRxiv (2016)
6. Hayer, K.E., Pizarro, A., Lahens, N.F., Hogenesch, J.B., Grant, G.R.: Benchmark analysis of algorithms for determining and quantifying full-length mRNA splice forms from RNA-seq data. Bioinformatics **31**, 3938–3945 (2015). https://doi.org/10.1093/bioinformatics/btv488
7. Kececioglu, J.D., Myers, E.W.: Combinatorial algorithms for DNA sequence assembly. Algorithmica **13**(1–2), 7–51 (1995)
8. Frazier, P.I.: A tutorial on Bayesian optimization. arXiv preprint arXiv:1807.02811, 8 July 2018
9. Bergstra, J., Bengio, Y.: Random search for hyper-parameter optimization. J. Mach. Learn. Res. **13**(Feb), 281–305 (2012)
10. Snoek, J., et al.: Scalable Bayesian optimization using deep neural networks. In: International Conference on Machine Learning, pp. 2171–2180, 1 June 2015
11. Snoek, J., Larochelle, H., Adams, R.P.: Practical Bayesian optimization of machine learning algorithms. In: Proceedings of the 25th International Conference on Neural Information Processing Systems - Volume 2, NIPS 2012, USA, pp. 2951–2959. Curran Associates Inc. (2012)
12. Frazier, P.I.: A tutorial on Bayesian optimization (2018)
13. Wang, L., Feng, M., Zhou, B., Xiang, B., Mahadevan, S.: Efficient hyper-parameter optimization for NLP applications. In: Proceedings of the 2015 Conference on Empirical Methods in Natural Language Processing, pp. 2112–2117 (2015)

14. Brochu, E., Cora, V.M., De Freitas, N.: A tutorial on Bayesian optimization of expensive cost functions, with application to active user modeling and hierarchical reinforcement learning. arXiv preprint arXiv:1012.2599, 12 December 2010
15. Jiang, Y., Kim, H., Asnani, H., Kannan, S., Oh, S., Viswanath, P.: LEARN codes: inventing low-latency codes via recurrent neural networks. arXiv preprint arXiv:1811.12707, 30 November 2018
16. The GPyOpt authors. GPyOpt: A Bayesian optimization framework in python (2016). http://github.com/SheffieldML/GPyOpt
17. Quitadadmo, A., Johnson, J., Shi, X.: Bayesian hyperparameter optimization for machine learning based eQTL analysis. In: Proceedings of the 8th ACM International Conference on Bioinformatics, Computational Biology, and Health Informatics - ACM-BCB 2017. ACM Press (2017)
18. Au, K.F., et al.: Characterization of the human ESC transcriptome by hybrid sequencing. Proc. Natl. Acad. Sci. U.S.A. **110**(50), E4821–4830 (2013)
19. Tilgner, H., Grubert, F., Sharon, D., Snyder, M.P.: Defining a personal, allele-specific, and single-molecule long-read transcriptome. Proc. Natl. Acad. Sci. U.S.A. **111**(27), 9869–9874 (2014)
20. Li, B., Dewey, C.N.: RSEM: accurate transcript quantification from RNA-seq data with or without a reference genome. BMC Bioinform. **12**(1), 323 (2011)

# Author Index

Printed in the United States
By Bookmasters